Sustainable Future for Human Security

Benjamin McLellan
Editor

Sustainable Future for Human Security

Environment and Resources

Springer

Editor
Benjamin McLellan
Graduate School of Energy Science
Kyoto University
Kyoto, Japan

ISBN 978-981-10-5429-7 ISBN 978-981-10-5430-3 (eBook)
https://doi.org/10.1007/978-981-10-5430-3

Library of Congress Control Number: 2017954387

Printed on acid-free paper

This Springer imprint is published by Springer Nature
The registered company is Springer Nature Singapore Pte Ltd.
The registered company address is: 152 Beach Road, #21-01/04 Gateway East, Singapore 189721, Singapore

Preface

This volume, *Sustainable Future for Human Security: Environment and Resources*, is the second of two in a series discussing a variety of critical issues for a sustainable and secure future for humanity. Sustainability is a systemic concern that can be examined from a variety of perspectives, at various levels of socio-environmental systems and sub-systems. Sustainable development is also a highly contextual concept, with no two societies or environments being exactly identical with regards to both endogenous factors and exogenous influences. It is therefore impossible to make a perfectly comprehensive examination of the topic of sustainability when considering its applications in (or interpretations from) the real world. However, a range of examples from a variety of fields of examination, such as that offered in these two volumes, should help to create an understanding of the broad landscape of sustainability.

This volume specifically presents on topics of natural resource conservation, environmental control, biology in tropical regimes, marine environments and food security. Such issues relate directly to the ability of "future generations to meet their own needs", as one of the key defining elements of sustainability. How do we manage resources as a society? How can we maintain the ability to provision society? Such questions are of high importance to sustainable development, but vary contextually.

This volume has a focus on tropical environments, which are vital to the mitigation of climate change and home to rapidly expanding societies. The chapters on natural resource conservation here are focused on agriculture and forestry, examining the rehabilitation of degraded land and the improved modelling of forest, as well as the linkages of agriculture and communities. The chapter on environmental control focusses on termites in this case, with a number of alternative perspectives on termite colonies. The chapter on biology and microbiology in tropical regimes covers a number of issues, largely dealing with the problems of waste valorization – for example, waste water treatment and biogas from agricultural waste.

The chapters on marine environments cover a variety of industrial-governance, aquaculture and waste impact remediation. Marine environments and resources have been increasingly important to global sustainable development, with treatment of these "commons" becoming increasingly prone to conflict. The degradation of marine environments, remediation and methods of governance are therefore vital, and have been incorporated in the United Nations Sustainable Development Goals.

The final section of this volume presents on food security and technology – covering social, political, environmental and technological components. Food security is again vital – predominately in developing countries, but in the long term becoming critical at a global scale.

The chapters presented in this volume were developed by authors who presented at the SUSTAIN 2015 conference, and have been reviewed by the conference committee. The editor would like to acknowledge the efforts of the authors, the editorial staff and the Sustain Society for the successful publication of this volume.

Kyoto, Japan Benjamin McLellan

Contents

Part I Natural Resource Conservation

1 **Revegetation Prospects on Ex-illegal Gold Mining Sites Using Secondary Forest Seedling Materials** 3
Wiwik Ekyastuti and Emi Roslinda

2 **Hydrological Processes in Different Types of Teak (*Tectona grandis* L.) Plantation** 11
Hatma Suryatmojo and M. Ali Imron

3 **Sustainability Assessment of Highland Tropical Rainforest Ecosystem: A Case of Selupu Rejang Agropolitan Area, Indonesia** 21
Sri Fatkhiati and Djoko Hartoyo

Part II Environmental Control

4 **The Diversity of Soil Fungus in and Around Termite Mounds of *Globitermes sulphureus* (Haviland) (Blattodea: Termitidae) and Response of Subterranean Termite to Fungi** 37
Ikhsan Guswenrivo, Hideyuki Nagao, and Chow Yang Lee

5 **Screening of Strains and Media of *Termitomyces eurrhizus* Collected from Japan Toward Nutritional Analyses** 53
Kazuko Ono, Tsuyoshi Yoshimura, and Kazuhiko Kinjo

6 **Volume Visualization of Hidden Gallery System of Drywood Termite Using Computed Tomography: A New Approach on Monitoring of Termite Infestation** 61
S. Khoirul Himmi, Tsuyoshi Yoshimura, Yoshiyuki Yanase, Toshiyuki Torigoe, Masanori Akada, Motoko Ikeda, and Setsuo Imazu

Part III Biomass and Microbiology in Tropical Regimes

7 The Potency of Fenton-Polyaluminum Chloride for Black Liquor Treatment 71
Ajeng Arum Sari, Anis Kristiani, Hendris Hendarsyah Kurniawan, and R. Irni Fitria Anggraini

8 Pretreatment of Sweet Sorghum Bagasse Using EFB-Based Black Liquor for Ethanol Production 85
Muryanto and Ajeng Arum Sari

9 Increasing Farmers' Affordability for Anaerobic Digester Construction in Rural Area Using Spatial Cluster Analysis 97
Christia Meidiana, Siska Ita Selvia, and Dian Dinanti

Part IV Marine Ecosystems

10 Marine Ecosystem Sustainability Post-Mine Closure Activities (Macrobenthos Dynamics Study Due to the Influence of Tailing Disposal in Buyat Bay, Minahasa, Indonesia) 115
Djoko Hartoyo, A. Harsono Soepardjo, Abimanyu T. Alamsyah, and Arie Herlambang

11 Reinventing Fiscal Policy on a Quintuple Helix Perspective Toward Indonesia-World Maritime Interaction: A Case in Batam Free Trade Zone and Lamongan, Indonesia 127
Haula Rosdiana, Inayati, and Maria R.U.D. Tambunan

12 Seaweed Utilization for Phytoremediation of *Litopenaeus vannamei* Shrimp Farming Waste in Recirculation Systems (Environmentally Friendly Design of Sustainable Shrimp Culture) 145
Sinung Rahardjo, A. Harsono Soepardjo, D. Djokosetiyanto, and Abimanyu T. Alamsyah

Part V Food Security and Technology

13 Effect of Salt Stress on Growth Performance and Antioxidant Status of Local Black Rice (*Oryza sativa L.*) 161
Yuda C. Hariadi, Arry Y. Nurhayati, and Rida R. Akbar

14 Larvicidal Activity of *Cerbera odollam* Gaertn Against a Dengue Vector, *Aedes aegypti* (Diptera: Culicidae) 175
Didi Tarmadi, Dwi Jayanti Gunandini, and Sulaeman Yusuf

15 DNA-Based Occupancy Modeling: Human Activity Did Not Significantly Influence Common Palm Civet's Occupancy 189
Sena A. Subrata, T.R.A. Budiman, and Fauzi Hamdan

16 Sustainable Development of Traditional Processed Fishery Products: A Feasibility and Risk Analysis of Small-Scale Businesses in West Kalimantan 197
Eva Dolorosa and Nurliza Nurliza

17 Obstacles to Achieving Food Security: The Failure of Rice Sector Policy and Its Impact on Peasant Deprivation in the Era of Political Reform in Indonesia 213
Sunardi Purwaatmoko

18 Identification of Phenotype and the Origin of Rambon Cattle of Indonesia 233
Yudi Adinata, Aryogi, Dicky Pamungkas, and Tety Hartatik

19 Machine Vision-Based Analysis for Black Tea Quality Evaluation 243
Avicienna Ulhaq Muqodas, Slamet Widodo, Kudang B. Seminar, and Mohamad Solahudin

Part I
Natural Resource Conservation

Chapter 1
Revegetation Prospects on Ex-illegal Gold Mining Sites Using Secondary Forest Seedling Materials

Wiwik Ekyastuti and Emi Roslinda

Abstract Illegal gold mining activities have been increasing over the last 20 years in West Kalimantan, including in Landak district. Illegal gold mining has caused environmental damage that needs to be reclaimed and revegetated. The research goal of this study was to review and explore the prospects of revegetation of the tailing areas of ex-illegal gold mining, using secondary forest seedling materials. Vegetation analysis in secondary forest was used with the line plots methods. Plots were systematically placed in two secondary forest locations (Mandor and Menjalin). In order to understand the species of plants that were relevant for revegetation in the ex-gold mining area, research was conducted by a census of plants in the ex-illegal gold mining area that had a succession. The results showed that in both sites of secondary forest, there was a potential to provide seeds for revegetation. In Mandor secondary forests, ten species of seeds were found, while in Menjalin secondary forests, 18 species of seeds were shown to have good availability (> 1000 seedlings per hectare). Among these plant species, nine species, ubah (*Syzygium zeylanicum*), laban (*Vitex pinnata*), bingir (*Ploiarium alternifolium*), cempedak (*Artocarpus heterophyllus*), pulai (*Alstonia pneumatophora*), medang (*Cinnamomum porrectum*), simpur (*Dillenia suffruticosa*), karet (*Hevea brasiliensis*), and jambu-jambuan (*Syzygium* sp.), were also found in the tailing areas of ex-illegal gold mining that have a succession. This means that the prospect of revegetation in the tailing areas of ex-illegal gold mining using secondary forest seedling materials is very good.

W. Ekyastuti (✉) • E. Roslinda
Faculty of Forestry, Tanjungpura University, Jl. Imam Bonjol, Pontianak 78124, Indonesia
e-mail: wieky_serdam@yahoo.co.id

B. McLellan (ed.), *Sustainable Future for Human Security*,
https://doi.org/10.1007/978-981-10-5430-3_1

1.1 Introduction

Illegal gold mining activities have been increasing over the last 20 years in West Kalimantan, including in Landak district. This activity has caused environmental damage, because the area has been left open (without vegetation), arid, nutrient poor, and polluted with heavy metals, especially mercury. The content of mercury in the tailing areas of illegal gold mining in Mandor, Landak district, was found to be 16 times higher than the permitted terrestrial threshold values (Ekyastuti 2013) and 2455 times higher than the permitted threshold values in aquatic sediments (Ekamawanti et al. 2005). This condition is very worrying, so it is apparent that the land needs to be reclaimed and revegetated. Reclamation activities can be done by improving the physical and chemical properties of the tailings. These chemical and physical improvements can be done with the addition of topsoil, organic matter, and fertilizer (organic fertilizer highly recommended). The recommended type of organic matter is compost. Compost should be given twice, at the time in the nursery with a ratio of 1: 1 (v/v) (Ekamawanti and Ekyastuti 2010) and at the time of planting in the area of the tailings (in the planting hole). Meanwhile, revegetation activities require species that are fast growing, adaptive, and catalytic (Setiadi 2003). Catalytic species have the nature of being fast growing, producing a lot of litter that is quickly decomposed and producing fruit that can invite the animals that play a role in seed dispersal so that there will be accelerated colonization of the area (Setiadi 2003).

One of the strategies for revegetation is restoration. Restoration is an effort to repair or restore the condition of the damaged area by forming the structure and function close to the original condition (Project RECA 2014). Most of the illegal gold mining in the Landak district has been done in forest areas. In revegetation which aims to achieve a condition of the structure and function of the area as before, it would require the plant species to also match the previous conditions. Therefore, the use of indigenous species is highly recommended. On the other hand, in Landak district, there are several secondary forests that are very high potential as a source of seedlings for revegetation in the tailing areas of ex-illegal gold mining.

The goal of this study was to review and explore the prospects of revegetation of the tailing areas of ex-illegal gold mining, using secondary forest seedling materials. The utilization of indigenous species in the revegetation program is very beneficial, because it is not only easy to get the seedlings but also to ensure the success of the growth.

1.2 Materials and Methods

The study was conducted in 2014 at two locations of secondary forests in Landak district, i.e., Mandor and Menjalin. A map showing the locations of the research is presented in Fig. 1.1. The method for vegetation analysis in secondary forest used

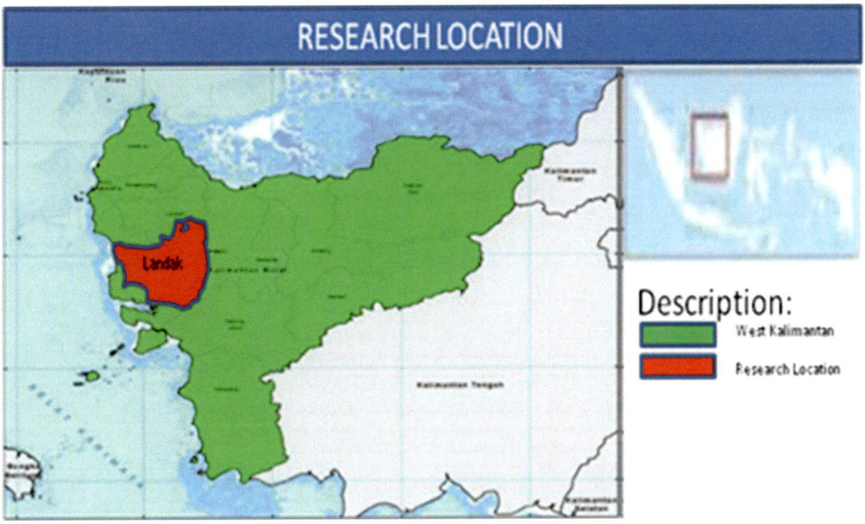

Fig. 1.1 Research location

line plots. Sample plot size of 20 m × 20 m were used to collect the data type of trees, plots of size 10 m × 10 m were used to collect the data type of the pole, plots of size of 5 m × 5 m sample were used to collect the data types of saplings, and a plot size of 2 m × 2 m samples were used to collect the data types of seedlings. As comparative data, vegetation analysis was also conducted in the tailing areas of ex-illegal gold mining using a census method to record the species of plants which grow well at that site. Plant census was conducted at the tailing areas of ex-illegal gold mining that have been abandoned >10 years. At these locations, two plots each measuring 0.5 hectares were made (Ekyastuti and Roslinda 2015). Furthermore, the data were analyzed by calculating the basal area, density, relative density, frequency, relative frequency, dominance, relative dominance, and importance value index (IVI) (Kusmana 1997).

1.3 Results and Discussion

1.3.1 Potential Secondary Forest as a Seedling Source

The results showed that the condition of secondary forests both in Mandor and in Menjalin were good and normal. Normal conditions were reflected from the uneven distribution of age classes (Supratman 2009); the vegetation was found in all age levels of seedlings, saplings, poles, and trees. In general, the density of plants per hectare decreased with increasing age of the plant. A graph showing the relationship between the number of plants per hectare with the phase of plant growth

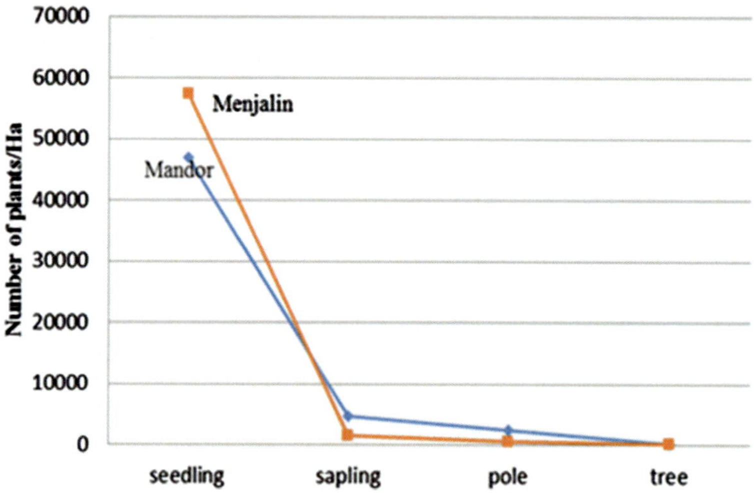

Fig. 1.2 Graphs of stand structure in secondary forests of Mandor and Menjalin

(diameter class) shows an inverted "J" pattern (Fig. 1.2). This indicates that the condition of secondary forests at both locations (Mandor and Menjalin) is normal (Meyer et al. 1961).

Based on the results of further investigation, it was known that the number of trees, poles, and saplings were 7478 per hectare in Mandor secondary forests and 2384 per hectare in Menjalin secondary forests. Both secondary forests are categorized as dense forest. This condition is dense forest because the forest covered more than 700 stems per hectare. Normal and dense forests illustrate that both secondary forests are healthy forests to serve as seed sources.

The good condition of the secondary forest was also indicated by the good natural regeneration. Good natural regeneration is the regeneration which has a density of seedlings of a species >1000 individuals/hectare (Syahri). In Mandor secondary forest sites, there are ten species of plants with a density > 1000 individuals/ha, namely, merkubung (*Macaranga gigantea*), simpur (*Dillenia suffruticosa*), bingir (*Ploiarium alternifolium*), pakeng (*Ficus* sp.), porang (*Amorphophallus oncophyllus*), pulai (*Alstonia pneumatophora*), ubah (*Syzygium zeylanicum*), kamayo, laban (*Vitex pinnata*), and medang (*Cinnamomum porrectum*). Meanwhile, in the Menjalin secondary forest sites, there are 18 species of plants with a density > 1000 individuals/ha, namely, amplasan (*Ficus ampelas*), bingir (*Ploiarium alternifolium*), cempedak (*Artocarpus integer*), jambu monyet (*Anacardium occidentale*), jambu-jambuan (*Syzygium* sp.), karet (*Hevea brasiliensis*), medang (*Cinnamomum porrectum*), medang ketingek (*Schima* sp.), meranti merah (*Shorea leprosula*), meranti putih (*Shorea* sp.), nilas, nilas hitam, pehengan, pendarahan, pulai (*Alstonia pneumatophora*), tarap (*Artocarpus odoratissimus*), ubah (*Syzygium zeylanicum*), and ubah merah (*Eugenia* sp.).

Table 1.1 Species of plants found in the tailing areas of ex-illegal gold mining

No.	Local names	Latin names	Family	Amount of individuals
1.	Simpur	*Dillenia suffruticosa*	Dilleniaceae	135
2.	Perepat	*Sonneratia alba*	Lythraceae	21
3.	Medang	*Cinnamomum porrectum*	Lauraceae	47
4.	Pulai	*Alstonia pneumatophora*	Apocynaceae	67
5.	Cempedak	*Artocarpus integer*	Moraceae	23
6.	Bingir	*Ploiarium alternifolium*	Theaceae	122
7.	Jambu-jambuan	*Syzygium sp.*	Myrtaceae	94
8.	Mahang merah	*Macaranga triloba*	Euphorbiaceae	12
9.	Leban	*Vitex pinnata*	Verbenaceae	35
10.	Karet	*Hevea brasiliensis*	Euphobiaceae	56
11	Ubah	*Syzygium zeylanicum*	Myrtaceae	72
12.	Nangka	*Artocarpus heterophyllus*	Moraceae	19
	Total amount of individuals			703

These data indicate that both locations have the potential as a source of seedlings for revegetation of tailing areas of ex-illegal gold mining.

1.3.2 Plant Species Found in the Tailing Areas of Ex-illegal Gold Mining

To identify the species that can grow well in the tailing areas of ex-illegal gold mining, we conducted a census of plants at that location. The census found 12 species of plants with varying amounts (Table 1.1). In general, plant conditions are generally good and healthy; there are no symptoms of mercury poisoning. This suggests that these species are very adaptive, easy to grow, and tolerant of the conditions and should be assumed to be phytoremediators of pollutants (in this case is mercury) (Setiadi 2003; Mansur 2010; Sarma 2011). This indicates that these plants would be appropriate to be planted in the tailing areas of ex-illegal gold mining.

In the ex-illegal gold mining areas, plant growth was found to be only on the level of seedlings, saplings, and poles and still dominated by seedlings and saplings. In this area plant growth at tree level was not found. This suggests that the process of succession in this area is still at an early stage. The succession process started from pioneer vegetation or early growth rate (Baasch et al. 2008). Over time, after the ex-gold mining is covered by crops, the pH will go up, so that the other plant species will begin to grow. Seen from the dominance of plants that are still seedlings and saplings in the tailings age $>$ 10 years after being abandoned, the speed of the succession in the tailings ex-illegal gold mining is indicated as being slow. Succession can be slightly accelerated at the beginning of the process by growing *Melastoma malabathricum*. This is because these plants can improve the nutrient content of the soil by reducing soil acidity (Suhardi 2008). In the study site, *M. malabathricum* was found only in a few places, generally on the riverbank (small rivers are formed because of the mining process). This situation led to the acceleration of the succession not being optimal.

1.3.3 Revegetation Prospects in the Tailing Areas of Ex-illegal Mining Using Secondary Forests Seedling Materials

Based on the data, there are nine species of plants that were found in the secondary forest sites (with a density > 1000 individuals/ha) and also found in the tailing areas of ex-illegal gold mining. These species are ubah (*Syzygium zeylanicum*), laban (*Vitex pinnata*), bingir (*Ploiarium alternifolium*), cempedak (*Artocarpus heterophyllus*), pulai (*Alstonia pneumatophora*), medang (*Cinnamomum porrectum*), simpur (*Dillenia suffruticosa*), karet (*Hevea brasiliensis*), and jambu-jambuan (*Syzygium* sp.). This shows that nine species of plants can be used for revegetation plants in the tailing areas of ex-illegal gold mining. It means that the prospect of revegetation in the tailing areas of ex-illegal gold mining using secondary forest seedling materials is very good.

The use of indigenous species, especially woody plants (besides undergrowth and shrubs) such as *Alstonia scholaris*, *Anacardium occidentale*, and *Eugenia* spp., is highly recommended (Tjhiaw and Djohan 2009; Holl et al. 2010). This is due to the planting of indigenous species being more likely to guarantee success in the restoration of mined land. In addition to the selection of appropriate species (preferably indigenous species) to further support the successful restoration of mined lands, some efforts need to be made, namely, improved soil before planting, application of appropriate silvicultural techniques, and the use of biological fertilizers (such as arbuscular mycorrhizal fungi).

1.4 Conclusion

The conclusion that can be derived from this study is that the prospect of revegetation of the tailing areas of ex-illegal gold mining in the Landak district is very good using secondary forest seedling materials. Through the utilization of indigenous species for revegetation in tailing areas of ex-illegal gold mining, its success can be optimized.

Acknowledgments We gratefully acknowledge the directorate general of Higher Education on the funding for this research through the national priorities research, Master Plan for Acceleration and Expansion of Indonesia Economic Development 2011–2025.

References

Baasch A, Tischew S, Bruelheide H (2008) Long-term monitoring of sandy dry grassland in a post-mining landscape. 6th European Conference on Ecological Restoration. Ghent, Belgium

Ekamawanti HA, Ekyastuti W (2010) Uji Efektivitas Isolat-isolat Mikrob Rizosfer Terhadap Pertumbuhan Beberapa Jenis Tanaman di Media Tailing yang Tercemar Merkuri. Laporan akhir Hibah Kompetitif Penelitian Sesuai Prioritas Nasional. Fakultas Kehutanan UNTAN; Pontianak

Ekamawanti HA, Utomo RS, Liwono (2005) Restorasi teresterial, riparian dan perairan areal bekas penambangan emas dengan teknologi bioremidiasi di Kecamatan Mandor Kabupaten Landak Kalbar. Laporan Akhir Penelitian Kegiatan Insentif Riset Matematika dan Ilmu Pengetahuan Alam; Fakultas Kehutanan UNTAN; Pontianak

Ekyastuti W (2013) Akselerasi suksesi vegetasi untuk mitigasi cemaran merkuri akibat penambangan emas skala kecil. Disertasi Fakultas Kehutanan UGM; Yogyakarta

Ekyastuti W, Roslinda E (2015) The potency of indigenous species as a mercury phytoremediator on illegal ex-gold mining reclamation. In: Proceedings of the International Symposium of IWoRS; organized by University of Sumatera Utara and Indonesian Wood Research Society (IWORS). ISSN:2459-9867. pp 156–161

Holl KD, Zipper CE, Burger JA (2010) Recovery of native plant communities after mining. Virginia Cooperative Extension Publication. http://pubs.ext.vt.edu/460/460-140/460-140.html

Kusmana C (1997) Metode survey vegetasi. Institut Pertanian Bogor, Bogor

Mansur I (2010) *Teknik silvikultur untuk reklamasi lahan bekas tambang*. Seameo Biotrop, Bogor

Meyer HA, Recknegel AB, Stevenson DD, Bartoo RA (1961) Forest management, 2nd edn. Ronald Press, New York

Project RECA (2014) Pedoman tata cara restorasi di kawasan konservasi. Project on Capacity Building for Restoration of Ecosystems in Conservation Areas; JICA dan PHKA: Jakarta

Sarma H (2011) Metal hyperaccumulation in plant; a review focusing on phytoremediation technology. J Environ Sci Technol 4(2):118–138

Setiadi Y (2003) Mycorrhizal inoculum production technique for land rehabilitation. Jurnal Manajemen Hutan Tropika 8(1):52–64

Suhardi (2008) The role of Melastoma malabathricum L. As alumunium acumulator plant to phosphorus availability on Ultisol Soil. Fakultas Pertanian UNIB, Bengkulu

Supratman SA (2009) *Manajemen Hutan*. Laboratorium Kebijakan dan Kewirausahaan Kehutanan; Fakultas Kehutanan UNHAS

Syahri K. Syarat kecukupan permudaan secara ekologis. http://khaerlita.syahri10.student.ipb.ac.id/2013/10/10/syarat-kecukupan-permudaan-secara-ekologis/

Tjhiaw G, Djohan TS (2009) Suksesi Vegetasi Alami di Bekas Tambang Timah Pulau Bangka. J Manusia dan Lingkungan 16(1):23–41

Chapter 2
Hydrological Processes in Different Types of Teak (*Tectona grandis* L.) Plantation

Hatma Suryatmojo and M. Ali Imron

Abstract Improved knowledge of canopy processes is allowing better estimation of the influence of forests in hydrology for forest function. The societal demand for timber wood has increased rapidly in the last two decades. The best genetic selection of teak (*Tectona grandis* L.) has been found and produced with shoot cuttings or tissue cultures and planted in Central Java, Indonesia. Understanding the hydrological processes in a teak plantation is helpful to develop a strategy for sustainable forest management system. The aim of this study was to measure the hydrological processes of canopy interception and direct runoff to estimate the forest interception in two types of the teak plantation. Two paired catchment experiments were established to monitor the canopy interception and direct runoff during the rainy season. Catchment T1 was planted with 10-year-old superior teak, and catchment T2 was planted with 10-year-old conventional teak. Superior teak is the result of tree improvement from prospective clonals of teak. Superior teak is propagated by shoot cutting or tissue cultures, while the conventional teak is from seed production in seed orchards. Canopy interception was investigated using 15 units of throughfall and eight units of stemflow in both catchments. The two catchment experiments were equipped with sharp-crested 90^0 V-notch weirs and automatic water level recorders. The percentage ratio between stemflow and throughfall in the T1 and T2 catchments were 0.08:54.7 and 0.15:70.4 of rainfall. The canopy interceptions in the T1 and T2 catchments were 45.2% and 29.4% of rainfall. Direct runoff in the T1 and T2 catchments were 32.2% and 47.2% of rainfall. Forest interceptions as a function of canopy interception and forest floor interception in the T1 and T2 catchments were 58.2% and 33.1% of rainfall. The difference in forest canopy cover of 13% has led to changes in the canopy interception. This change has affected an increase in the direct runoff and forest interception in the catchment. Characteristics of superior teak with fast growth and

H. Suryatmojo (✉)
Laboratory of Watershed Management, Faculty of Forestry, Universitas Gadjah Mada, Jl. Agro No. 1 Bulaksumur, Yogyakarta 55281, Indonesia
e-mail: hsuryatmojo@ugm.ac.id

M.A. Imron
Laboratory of Wild Life Management, Faculty of Forestry, Universitas Gadjah Mada, Jl. Agro No. 1 Bulaksumur, Yogyakarta 55281, Indonesia

B. McLellan (ed.), *Sustainable Future for Human Security*,
https://doi.org/10.1007/978-981-10-5430-3_2

uniform physical appearance of diameter, height, and canopy structure showed higher forest interception than that of the conventional teak.

2.1 Introduction

In the forested area, movement of water between the atmosphere and the soil plays a diversified role in the water storage capacity. Hydrologic processes in the forested catchment start from rainfall interception in the forest vegetation structure. A portion of precipitation is inevitably intercepted by the canopy (canopy interception), flows along the stem to the ground surface (stemflow), drips from the foliage and branches or passes through canopy openings to the ground (throughfall), or is further intercepted at the forest floor (litter interception). These processes cause a reduction in precipitation quantity and a redistribution of precipitation toward the soil. Forest interception is an important event in the hydrologic cycle because of its effects on rainfall deposition, soil moisture distribution, wind movement, heat dissipation, and impact energy of raindrops on soil erosion (particle detachment) (Chang 2006).

The hydrologic cycle in the tropical rainforest is a unique system with interdependence between its components. Disruption of one component will impact on others and cause instability in the hydrologic cycle. Vegetation cover change has a profound influence on the hydrological cycle. Forest harvesting, or other reductions to vegetative cover, generally increases the average surface runoff volume and total water yield of a watershed. These vegetation cover alterations potentially decrease the time concentration of flow, increase the intensity of peak flows for a given precipitation event, and increase the frequency and intensity of extreme flow events, especially channel-forming flows. These alterations tend to deteriorate water quality by transporting sediment and other pollutants from the landscape and increasing erosive forces within the stream channel. Our previous studies have indicated that forest structures are a principal cause of hydrological differences between watersheds (Suryatmojo et al. 2014). Thus, an understanding of relationships between forest structure and runoff processes is essential for the quantitative prediction of the effects of deforestation and changes in vegetation (Bent 2001; Bosch and Hewlett 1982; Lanea and Mackay 2001).

Different land use practices affect the infiltration rate of soil in different ways, depending on their effects on the intrinsic properties of the soil(Osuji et al. 2010), and clear cutting and intensive plantation systems are suspected to dramatically impact soil and reduce the forest cover. The loss of forest cover from forest harvest reduces interception of raindrops (increasing drop impact energy and soil detachment) and reduces evapotranspiration, increasing the amount of water available for infiltration, soil storage, and runoff. Therefore, soil moisture capacity is reached with less rainfall, and any excess can produce surface runoff and increase peak flows and streamflow volumes. In a previous study, the infiltration capacity of a tropical rainforest 1 year after selective logging and intensive line planting treatment decreased to 81.8% of that of a virgin forest (Suryatmojo 2014). Many studies have indicated that forest vegetation covers are a principal cause of hydrological differences between catchments (Abdulhadi et al. 1981; Liu et al. 2006; Zhang et al. 1999; Zhao et al. 2009).

Teak (*Tectona grandis* L.) grows naturally throughout the Southeastern Asia and is one of the most valuable tropical hardwood species on the international market. In Indonesia, especially on the island of Java, teak grows in optimum at elevations of 100–700 m, annual rainfall of 1500–2500 mm, temperature of 27–36^0 C (noon) and 20–23^0 C (night), good in drainage, and clear difference between rainy and dry seasons. Teak trees attain a height of 39–45 m under such favorable conditions and can be harvested after 60 years. In Java island, there is a teak forest with an area of 1.2 million hectares, but the productive area is less than 500.000 hectares. The need of high production of teak was answered by the development of the superior teak. Superior teak is the result of tree improvement from prospective clonals of teak and propagated by shoot cutting or tissue cultures. It started being planted in 2005. In several plantation sites (Gunungkidul, Cepu, and Ngawi Districts), the growth performance of superior teak was excellent with average diameter growth of 3 cm/year and it was predicted to be ready to harvest in 15–20 years. The difference of growth performance influenced the hydrological processes at the catchment level. The aim of this research was to measure the forest interception between catchments with different types of teak canopy.

2.2 Methods

2.2.1 Study Site

This research was conducted in a teak plantation forest of PERUM PERHUTANI, a semiprivate forest company. Two paired small catchment experiments were established in "Pasar Sore" forest area, Cepu District, Central Java. The two paired small catchments are planted with two different types of teak, superior teak (0.4 hectare) and conventional teak (0.3 hectare). Hereafter, these catchments are referred to as the "T1 catchment" and the "T2 catchment," respectively (Fig. 2.1).

2.2.2 Canopy Cover

Canopy cover density in the T1 and T2 catchments were measured in 20 × 20 m sample plots. Tree canopy cover profiles in the catchment are shown in Fig. 2.2.

Superior teak is the result of tree improvement from prospective clonals of teak. It is propagated by shoot cutting or tissue cultures, while the conventional teak is

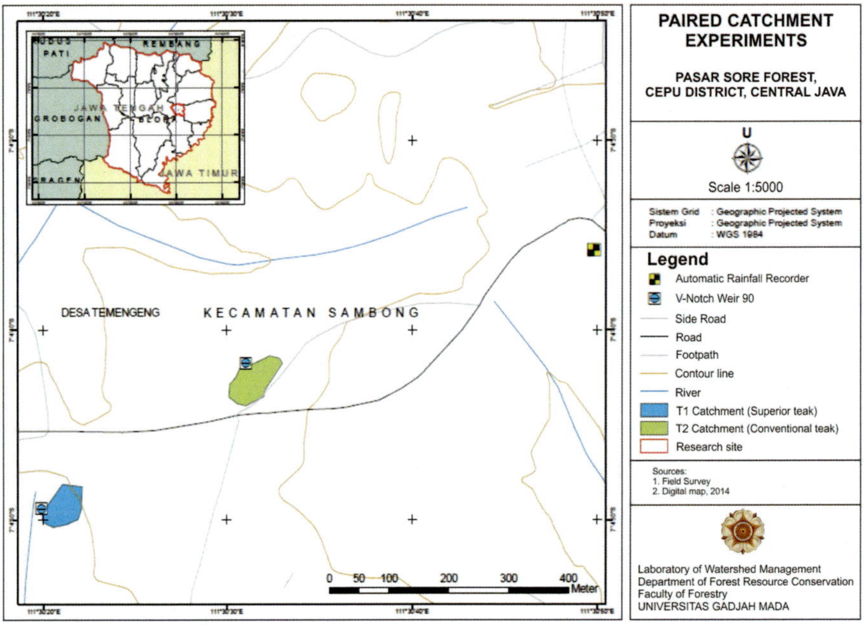

Fig. 2.1 Location of paired catchment experiment

propagated by seed production in seed orchards. Propagation using shoot cutting or tissue cultures will produce large numbers of seedlings at the same time, and these will have similar physical characteristics with the parent trees in growth performance such as diameter, height, canopy wide, and wood quality. Seed production in the seed orchards has a high variation of physical characteristics in the plantation areas. High genetic variation in the seed orchards caused different growth performances. The canopy cover conditions in both catchments are shown in Fig. 2.2. In the T1 catchment, the canopy cover density is higher than that in the T2 catchment. The percentage tree canopy cover was 85.12% in the T1 catchment, while in the T2 catchment was 75.37%. Thus canopy cover in the T1 catchment was 13% higher than that in the T2 catchment. High canopy cover density will lead to greater canopy interception and reduce the net precipitation in the forest floor. This will reduce the potential of direct runoff in the catchment.

2.3 Results and Discussion

2.3.1 *Canopy Interception*

Rainfall, stemflow, throughfall, and direct runoff were investigated from April to June 2015. Precipitation was monitored with an interval of 5 min using an

Fig. 2.2 Horizontal and vertical structures of canopy cover. (**a**) T1 catchment; (**b**) T2 catchment

automatic tipping bucket rain gauge located near the two catchments (Fig. 2.1). The discharge rates were measured using 90° V-notch weirs (0.6 × 0.8 × 0.45 m) and water level loggers (HOBO U-20) with a time interval of 5 min at each catchment outlet (Fig. 2.3a). During the research period, we selected an individual rainfall and single hydrograph data at each catchment. The hydrograph was analyzed the direct runoff volume for each hydrograph. The canopy interception in each catchment was investigated using 15 throughfall gauge units and 8 stemflow gauge units (Fig. 2.3b).

Forest interception (I_F) is the sum of canopy interception (I_C) and forest floor interception (I_{FF}) using Eq. (2.1):

$$I_F = I_C + I_{FF} \quad (2.1)$$

Net precipitation (P_N) is the sum of throughfall (P_{TH}) and stemflow (P_S) using Eq. (2.2):

$$P_N = P_{TH} + P_S \quad (2.2)$$

Fig. 2.3 (**a**) 90° V-notch weirs; (**b**) installation of stemflow and throughfall

Forest floor interception is calculated using Eq. (2.3):

$$I_{FF} = P_G - P_N - DRO \tag{2.3}$$

Canopy interception (I_C) is the difference between gross precipitation in the open (P_G) and net precipitation (P_N) using Eq. (2.4):

$$I_C = P_G - P_N = P_G - (P_{TH} + P_S) \tag{2.4}$$

Forest canopies serve as a barrier against precipitation reaching the ground. These processes cause a reduction in precipitation quantity and a redistribution of precipitation toward the soil. Canopy density influenced the capacity of rainfall storage. Stemflow and throughfall in the T2 catchment are higher than in the T1 catchment (Fig. 2.4). These indicate that the T1 catchment has higher rainfall storage capacity than that of the T2 catchment. The rainfall storage capacity in the T1 catchment is influenced by higher canopy density (Fig. 2.2).

Intensive plantation using superior teak in the T1 catchment has significantly changed the forest canopy cover. Variation of canopy cover density influenced the hydrologic process in runoff response. The differences in the canopy interception capacity are shown in Fig. 2.5 and Table 2.1.

Table 2.1 shows that the canopy interception in the T1 catchment was 1.5 times higher than that in the T2 catchment. The similarity of architectural canopy in the T1 catchment has increased the canopy capacity to catch and store rainfall more effectively.

2.3.2 *Direct Runoff*

The hydrologic behavior of small catchments tends to be different from that of large catchments. A small catchment is very sensitive to high-intensity rainfall of short duration and land cover characteristics (Chang 2006). To increase the understanding of runoff response between catchments, an individual rainfall event and a single

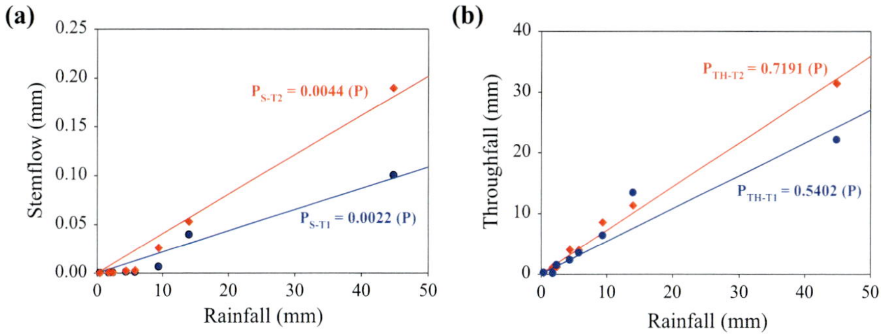

Fig. 2.4 Relationship between rainfall and stemflow (**a**) and throughfall (**b**)

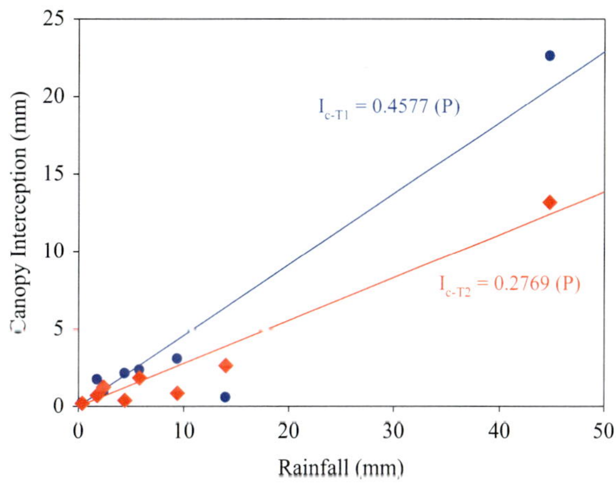

Fig. 2.5 Relationship between rainfall and canopy interception in the T1 and T2 catchments

Table 2.1 Percentage of canopy interception in the T1 and T2 catchments

Throughfall (P_{TH}), % of P_G		Stemflow (P_S), % of P_G		Net rainfall (P_N), % of P_G		Canopy interception (I_C), % of P_G	
T1	T2	T1	T2	T1	T2	T1	T2
54.7	70.4	0.08	0.15	54.70	70.56	45.22	29.44

runoff hydrograph at each catchment were selected during field investigation period to clarify the direct runoff response to rainfall events. The rainfall and direct runoff response analysis is shown in Fig. 2.6.

The results showed that the rainfall and direct runoff response in the T2 catchment was higher than that in the T1 catchment. The percentages of direct runoff and rainfall in the T1 and T2 catchments were 32.2% and 47.15%. Average direct runoff in the T2 catchment was 1.4 times higher than that in the T1 catchment. Land use change in a catchment may lead to changes in its water balance. The response time of stream flow is determined by climate (mostly

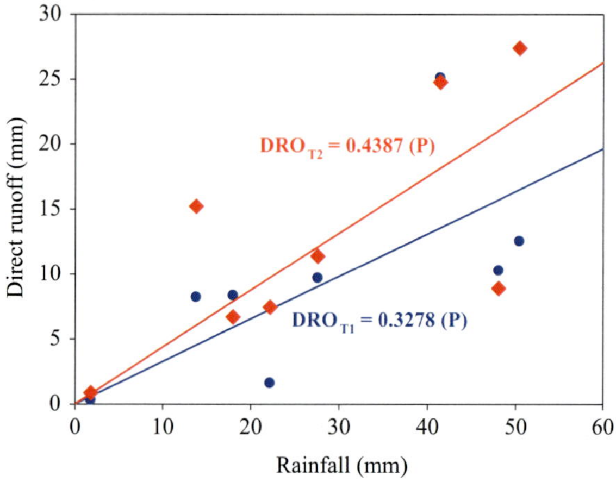

Fig. 2.6 Relationship between rainfall and direct runoff in the T1 and T2 catchments

Table 2.2 Forest interception in the two catchments

Canopy interception (I_C) % of P_G		Direct runoff (DRO) % of P_G		Forest interception (I_F) % of P_G	
T1	T2	T1	T2	T1	T2
45.22	29.44	32.21	47.15	58.23	33.04

rainfall), vegetation characteristics, catchment properties, and vegetation management practices (Suryatmojo et al. 2014). The forest canopy serves as a barrier against precipitation reaching the ground.

The growth of vegetation in the catchment increased canopy interception and forest floor retention. For small rainfall events, most precipitation was trapped by the canopy in the T1 catchment, leading to negligible direct runoff responses to rainfall (Fig. 2.6). Although the canopy cover in the T2 catchment was almost the same as that in the T1 catchment, canopy interception was slightly lower than that in the T1 catchment (Table 2.2). Lower canopy cover led to lower evapotranspiration and higher net precipitation in the T2 catchment. In the T2 catchment with a low canopy cover density, there was a large net precipitation and increased amounts of direct runoff (Fig. 2.6). The high canopy cover density in T1 catchment areas controlled the net precipitation by canopy interception. The relationship between rainfall and throughfall-stemflow is shown in Fig. 2.4, canopy interception in Fig. 2.5, and direct runoff in Fig. 2.6; forest interception among the two catchments is shown in Table 2.2.

The forest interceptions in the T1 catchment and T2 catchment were 58.23% and 33.04% of rainfall. The results indicate that the difference of architectural canopy cover in the teak plantation forest has significantly changed the hydrological processes in the catchment.

2.4 Conclusions

We investigated the hydrological responses of two types of teak forest plantation using two small catchments. The effect of forest management system on hydrological responses is an important factor in catchment hydrology and forest management. Our monitoring data provided strong linkages between architectural canopy cover and canopy interception and forest interception. We found evidence that differences in forest canopy cover of 13% have led to changes in canopy interception. This change has affected an increase in the direct runoff and forest interception in the catchment.

Forest managers should consider the impact of vegetation structure to the canopy interception, direct runoff, and forest interception. We recommend controlling canopy cover density and combining ecologically-based vegetation structure design would be an effective way to control the hydrological processes in the catchment.

References

Abdulhadi R, Kartawinata K, Sukardjo S (1981) Effects of mechanized logging in the lowland dipterocarps Forest at Lempake, East Kalimantan. The Malaysian Forester 44:407–418

Bent GC (2001) Effects of Forest-management activities on runoff components and ground-water recharge to Quabbin reservoir, Central Massachusetts. For Ecol Manag 143:115–129

Bosch JM, Hewlett JD (1982) A review of catchment experiments to determine the effect of vegetation changes on water yield and evapotranspiration. J Hydrol 55:3–23

Chang M (2006) Forest hydrology, an introduction to water and forests, Second edn. CRC Press, U.S.A., pp 201–210

Lanea PNJ, Mackay SM (2001) Streamflow response of mixed-species eucalypt forests to patch cutting and thinning treatments. For Ecol Manag 143:131–142

Liu Y, An S, Deng Z, Fan N, Yang H, Wang Z, Zhi Y, Zhou C, Liu S (2006) Effects of vegetation patterns on yields of the surface and subsurface waters in the Heishui Alpine Valley in west China. *Hydrol Earth Syst Sci Discuss* 3:1021–1043

Osuji GE, Okon MA, Chukwuma MC, Nwarie II (2010) Infiltration characteristics of soils under selected land use practices in Owerri, southeastern Nigeria. World Journal of Agricultural Sciences 6(3):322–326

Suryatmojo H (2014) Recovery of Forest soil disturbance in the intensive Forest management system. Procedia Environ Sci 20., Elsevier:832–840

Suryatmojo H, Fujimoto M, Kosugi K, Mizuyama T (2014) Runoff and soil erosion characteristics in different periods of an intensive forest management system in a tropical Indonesian rainforest. International Journal of Development and Planning 9(6):830–846

Zhang L, Dawes WR, Walker GR (1999) Predicting the effect of vegetation changes on catchment average water balance. Cooperative Research Centre for Catchment Hydrology. CSIRO Land and Water, Clayton

Zhao FF, Zhang L, Xu ZX (2009). Effects of vegetation cover change on streamflow at a range of spatial scales. In: 18th World IMACS/MODSIM Congress, Cairns, Australia 13–17 July 2009

Chapter 3
Sustainability Assessment of Highland Tropical Rainforest Ecosystem: A Case of Selupu Rejang Agropolitan Area, Indonesia

Sri Fatkhiati and Djoko Hartoyo

Abstract This study aimed to analyze the status and sustainability index of agropolitan areas in the highland tropical rainforest ecosystem. For the assessment of the sustainability status, this study uses the integration of a variety of operational indicators. Indicators were prepared consisting of rural development indicators, agropolitan performance indicators, and indicators of sustainable agriculture in the highland tropical rainforest ecosystem. The analysis used a modification of Rapfish by using multidimensional scaling method called RAP-Agrotropika. Results show that the status of multidimensional sustainability index is less sustainable. There are two dimensions that are less sustainable, namely, the environmental dimension (48.61) and the infrastructure and technology dimension (48.31). Dimensions of economic, social, and institutional sustainability status are quite sustainable with respective index values of 50.93, 50.08, and 56.71. Leverage attributes of each dimension were pesticides (environmental dimension); the proportion of employment in agricultural sector, development of farm-level price, and vegetable market (economic dimension); the level of knowledge about the environment (social dimension); rural community access to the centers of health and education services and the number of postharvest processing infrastructure (infrastructure and technology dimension); and the proportion of off-farm activities which were managed or owned by local communities (institutional dimension). To improve the sustainability status of the area, attention should be paid particularly to the leverage attributes.

S. Fatkhiati (✉) • D. Hartoyo
The Coordinating Ministry of Maritime Affairs, Republic Indonesia, Jl.M.H. Thamrin no. 8, Central Jakarta, Indonesia
e-mail: fettysri@gmail.com

B. McLellan (ed.), *Sustainable Future for Human Security*,
https://doi.org/10.1007/978-981-10-5430-3_3

3.1 Introduction

The concept of rural development that has been executed to date has apparently not been able to improve the welfare of farmers and the development of rural areas, tending to cause a gap between urban and rural regions. There is an imbalance between rural and urban development, which has driven the emergence of a development model that formally acknowledges the linkages between rural and urban areas. This model is implemented through the agropolitan concept of development of rural areas in order to improve the rural economy to the equivalent of the city.

Agropolitan is combining a regional approach and cross-sector integration in development. However agropolitan development is oriented to increase agricultural production, so that economic growth increases till it is a fast-growing region. Pressure arising on the environment as a result of intensive farming activity, especially in the highland agricultural areas, can cause the area to become unsustainable.

Agropolitan at the conceptual level has colored the implementation of development programs in Indonesia. Agropolitan in the sense of a real and planned program started in 2002 involving various related sectors. The agropolitan program has been implemented in several Asian countries including Indonesia, by setting eight pilot areas; one of them is Rejang Lebong district, Bengkulu Province.

Bengkulu Province has been established as a pioneering area of agropolitan development since 2002, i.e., the agropolitan area of Selupu Rejang. This agropolitan area has still not reached its ideal target. The disadvantage is that this concept is loaded with the economic development of a region to be rapidly growing and has not been studied in the context of sustainable development. The implementation is more focused on increasing production; as a result, it has not adequately addressed the means to reach the target of integrated regional development which includes the aspect of sustainability of the environment.

Rejang Lebong Regency is located on a plateau with a hilly topography. Farmers cultivate a variety of upland vegetables on hilly terraced land with varying degrees of slope. Agricultural land in the wet tropics is not easy to cultivate in a sustainable and profitable way. The way farming is done is basically counter to nature and ignores the wet tropical ecological rules. Agricultural cultivation that would take into account this environment would be developed as agricultural cultivation within tropical rainforest ecosystems.

Emerging pressures on the environment due to intensive cultivation activities, especially in highland agriculture, cause the region to become unsustainable. The data shows that environmental degradation has occurred in the agropolitan area of Selupu Rejang, Bengkulu Province – for example, dependence on high inputs, decreasing the quality of river water marked with turbidity, DO, BOD exceeding the water quality standard limits(Environment Agency of Bengkulu Province 2013), and erosion of 248.59 ton/Ha/year in the catchment of Lake Mas Harun

Bastari due to agricultural activity in the lake border area (Environmental Research Center of Bengkulu University 2006).

In addition to environmental degradation issues, the management of agropolitan Selupu Rejang area is still incomplete. Agropolitan development is still impressed as the development of agricultural sector in the broad sense, not the concept of integrated development of the region (The Agropolitan Infrastructure and Facilities Support Unit 2005; Agropolitan Working Group of Rejang Lebong Regency 2007). As a result, the added value expected from the development of this region has not been achieved. Increased production is not necessarily accompanied by increased welfare of vegetable farmers.

Agricultural development in the agropolitan concept should ideally be done from the perspective of rural development and in harmony with the environment. However, in both the agropolitan concept initiated by Friedmann and Douglass (Friedmann 1981), as well as the Indonesian agropolitan concept initiated by Rustiadi(Rustiadi 2007), and the Ministry of Agriculture(Ministry of Agriculture 2002), the concept has not given sufficient attention relating to aspects of sustainability. If the concept is not given sufficient attention in relation to environmental issues, the implementation of agropolitan over the long term becomes unsustainable.

Sustainable agricultural development seeks to achieve sustainability of agricultural production, rural economic sustainability, and long-term environmental sustainability (Li 2009). Based on the definition of sustainability, the sustainable management of agropolitan areas in this study is management that considers economic, social, and environmental aspects in an integrated manner in order to achieve sustainable agricultural production, rural economic sustainability, and environmental sustainability in accordance with the typical Indonesian ecosystem within the framework of development of rural areas.

Understanding the principles of linkage and mutual influence between the characteristics and components in the agropolitan area becomes very important. However, this has not been fully utilized in the management of agropolitan Selupu Rejang area. This is due to the lack of understanding of the linkages of regional characteristics and important components in the management of the area. Therefore, the analysis of the characteristics of the management of the area becomes the starting point to be able to study and analyze the sustainability of regional management.

Sustainability analysis becomes very important, as it will elucidate variables that affect sustainability in each dimension being assessed. An integrated approach is needed to analyze sustainability. In order for the concept of sustainability to be operationalized, widely conducted research to assess sustainability has used a variety of indicators, which are incorporated in various dimensions within the scope of the investigation.

To assess the status of sustainability, effective selection of indicators is key. Indicators should be chosen with care, so as to measure and explain clearly the conditions of sustainability. The criteria of indicators are to be easily measured, sensitive to pressure, anticipatory, able to predict changes, and integrative (Dale and Beyeler 2001).

A number of operational indicators are used in this study. These indicators include those relating to agroecosystems, pressure on the agroecosystem, the vulnerability of agroecosystems, and agroecosystem management (Friedmann 1981). Besides that, indicators are also compiled based on systemic attributes consisting of productivity, stability, reliability, resilience, and adaptability (Rao and Rogers 2006; Binder et al. 2010; Rodrigues et al. 2010; Bosh et al. 2012). Indicators also include operational indicators to measure the sustainability of agriculture in developing countries (Zhen and Routray 2003). Consideration of selection of indicators that represent each dimension of sustainability must be adapted to the characteristics and scope of the study. This study uses the integration of indicators or attributes that make up the dimensions of sustainability. Indicators or attributes that are used are therefore a combination of performance indicators in rural development, agropolitan performance indicators, and operational indicators of sustainable agriculture in the highland tropical rainforest ecosystem.

3.2 Material and Methods

3.2.1 Area Studied and Data Collected

The study was conducted in Selupu Rejang agropolitan region, Rejang Lebong Regency, Bengkulu Province (Fig. 3.1). The data collected were both primary and secondary data. Primary data were collected through surveys, interviews, and observations. Secondary data were collected from the results of previous research, reports related to service agencies, and the data from the Central Bureau of Statistics.

3.2.2 Analysis Methods

The analysis was performed with the approach of multidimensional scaling (MDS) modified from the method rapid assessment techniques for fisheries (RAPFISH). The analysis method is called RAP-Agrotropika. Attributes were compiled based on literature references, expert opinion, and analysis of the existing area. There are 62 attributes that make up the dimensions of agropolitan sustainability status. These consist of 11 attributes that make up the environmental dimension, 16 attributes compose economic dimension, 8 attributes make up social dimension, 17 attributes make up the social dimension, and 11 attributes make up the institutional dimension. The unit of analysis for the measurement of status of sustainability is a

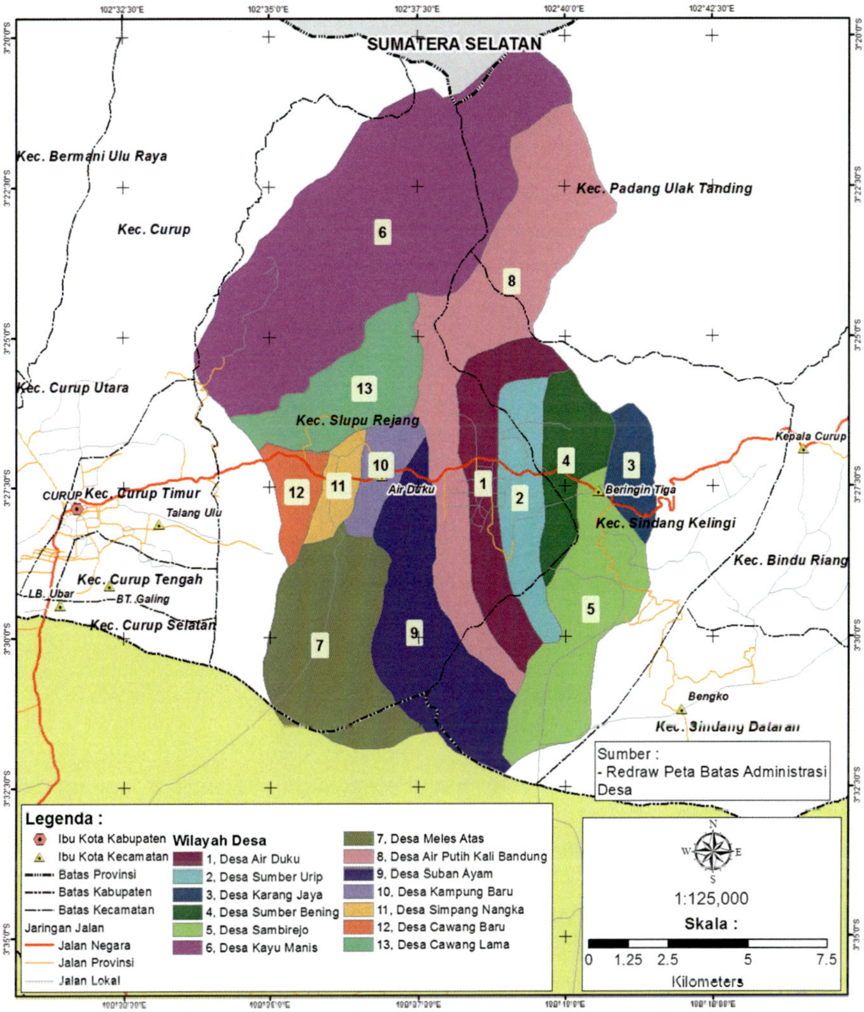

Fig. 3.1 Map of Selupu Rejang agropolitan area

subdistrict. Measurements are made by giving a score on the assessment criteria that have been compiled using a questionnaire with experts.

3.3 Results

Rap-Agrotropika analysis of the 11 attributes that make up the environmental dimension shows that the value of the sustainability index on the environmental dimension is 48.61. The values are on a scale from 25.01 to 50.00 (out of 100) with

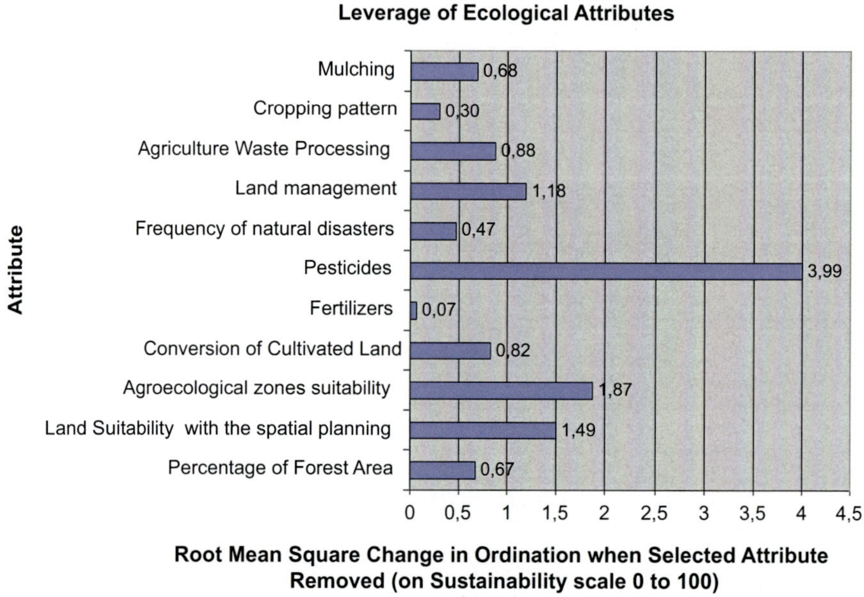

Fig. 3.2 Leverage attributes of environmental dimension

an overall status being less sustainable. Results of the leverage analysis of the attributes that make up the environmental dimension acquired one attribute as being the most sensitive. The attribute is the use of pesticides (3.99) as shown in Fig. 3.2.

The results show that in order to improve the sustainability status of the region, it is necessary to pay attention to the attributes which have value changes of root mean square (RMS) which is high. The use of pesticides is an attribute that has the most impact on the sustainability value of the environmental dimension. Based on the survey of this study, 89.0% of respondents never use organic pesticides to control pests and plant diseases. Expert assessment results also show that the use of pesticides is in a bad score range.

Vegetable farmers in Selupu Rejang generally applied pesticides less than 5 times a week; the rest apply between 6 and 15 times a week. In 1 day 46% of the farmers expressed using 3 containers of pesticides (Muktamar et al. 2012). Conditions of high pesticide use will cause reduced diversity. The increasing intensification of agriculture will also change soil conditions (Noordwijk and Hairiah 2006a). The more intensive agriculture is, the more likely that the biodiversity of soil organisms will decrease.

Rap-Agrotropika analysis of the 16 attributes that make up the economic dimension shows that sustainability index of economic dimension is 50.93. The values are on a scale from 50.01 to 75.00 with a status quite sustainable. Results of leverage analysis obtained three attributes as most sensitive. These attributes are the

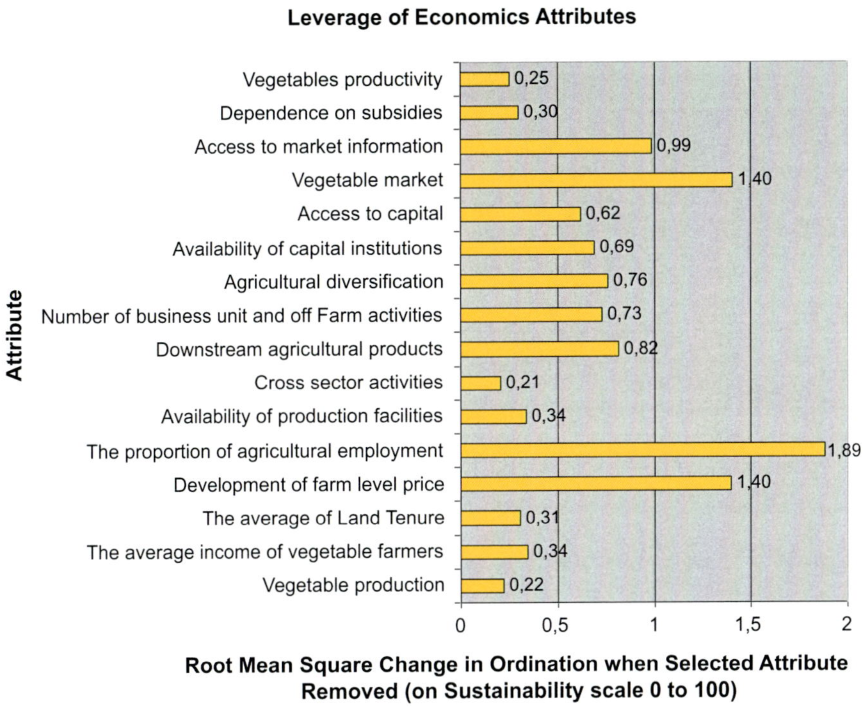

Fig. 3.3 Leverage attributes of economics dimension

proportion of agricultural employment (1.89), development of farm level price (1.40), and the vegetable market (1.40) as shown in Fig. 3.3.

More than 50% of the population depend on agriculture. In addition to the agricultural sector employment, development of farmer level pricing and markets of agricultural products should be paid attention for improving regional sustainability. Under existing conditions, the resulting output cannot provide added value when viewed from the aspect of the market, given the weakness of the product's ability to compete.

Rap-Agrotropika analysis of eight attributes that make up the social dimension shows that the value of the social dimension sustainability index is 50.08. The values are on a scale from 50.01 to 75.00 with a status quite sustainable. Leverage analysis results on the attributes that make up the social dimension show one attribute to be most sensitive, which is the level of knowledge, as shown in Fig. 3.4.

The level of knowledge about environment and education is very closely linked. Education is closely linked to the application of cultivation techniques and conservation techniques by farmers. Education is positively related to the perception of the problems of erosion and positive behavior on land conservation. With a good education, farmers will have the ability to improve the management of the farm also. Farmer education is correlated with the value of vegetable production.

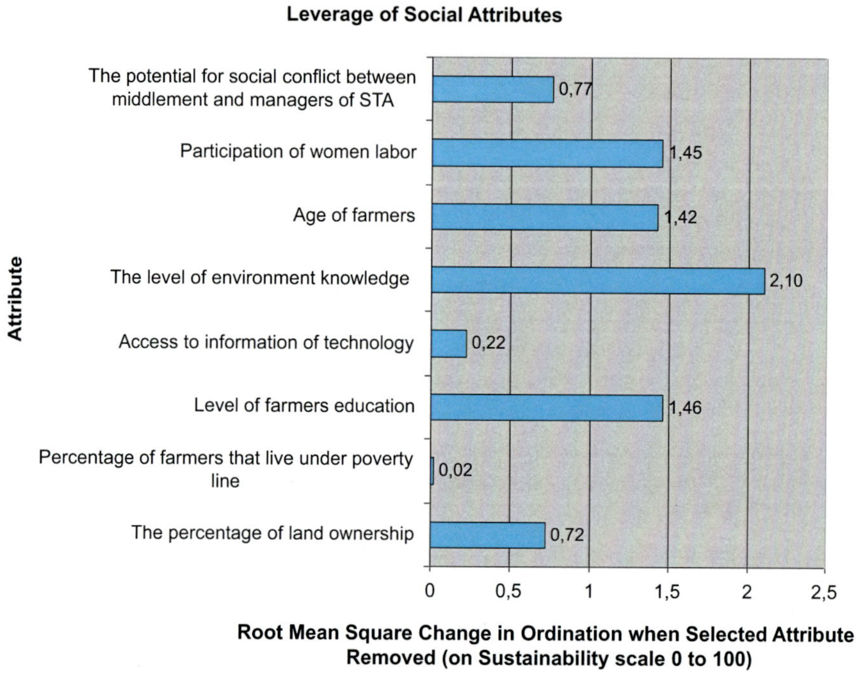

Fig. 3.4 Leverage attributes of social dimension

Improved education of farmers by 1% would increase production value of 0.54%, assuming other variables remain constant (Sumantri and Sukiyono 2011).

Rap-Agrotropika analysis of the 16 attributes that make up the infrastructure and technology dimension shows that the value of the infrastructure and technology dimension in the sustainability index is 48.31. The values are on a scale from 25.01 to 50.00 with a status of "less sustainable." Leverage analysis results on the attributes that make up the infrastructure and technology dimension show two attributes to be most sensitive. These attributes are the rural people's access to healthcare centers and education and number of postharvest processing infrastructure as shown in Fig. 3.5.

The leverage analysis showed that rural people's access to healthcare centers and education provided a strong influence on the sustainability of the agropolitan area. This attribute is very rarely noticed. Policy makers are more focused on improving infrastructure to support the production systems. The number of processing infrastructure also has a poor score, as does postharvest technology. Results of this analysis showed the need for the development of processing, in parallel with the dissemination of good postharvest technology.

Rap-Agrotropika analysis results on 11 attributes which make up the institutional dimension show that the institutional dimension sustainability index is 56.71. The values are on a scale from 50.01 to 75.00 with a status of "quite sustainable." Leverage analysis results on the attributes that make up the institutional dimension

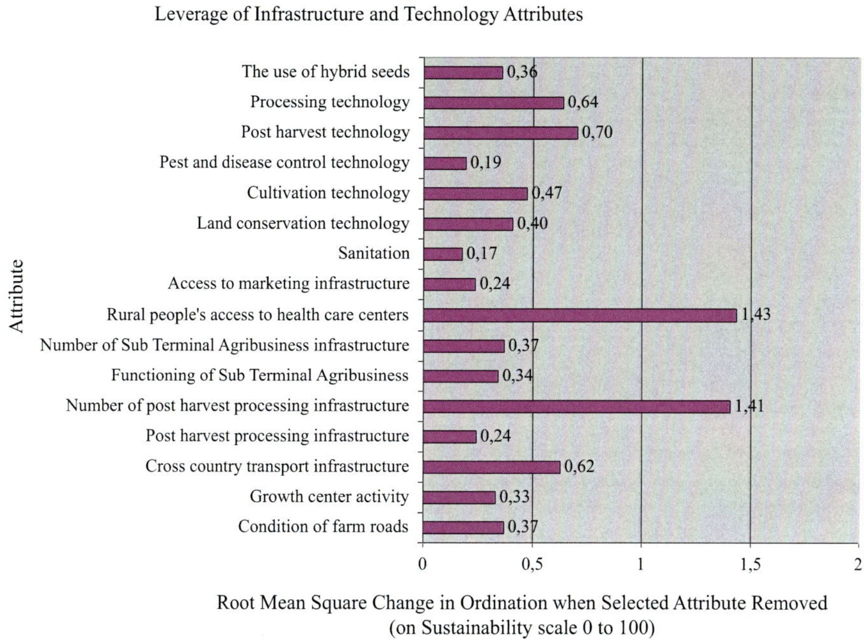

Fig. 3.5 Leverage attributes of infrastructure and technology dimension

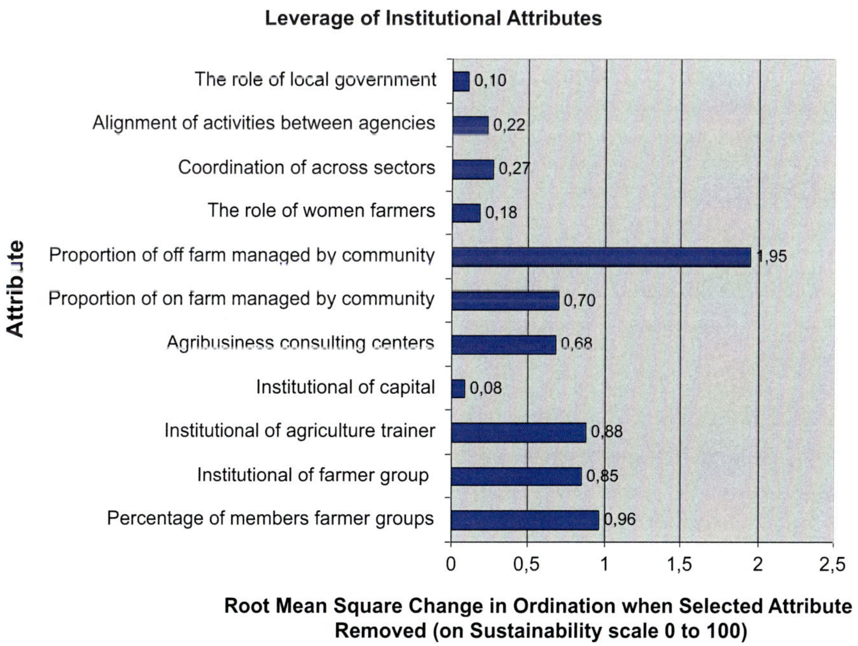

Fig. 3.6 Leverage attributes of institutional dimension

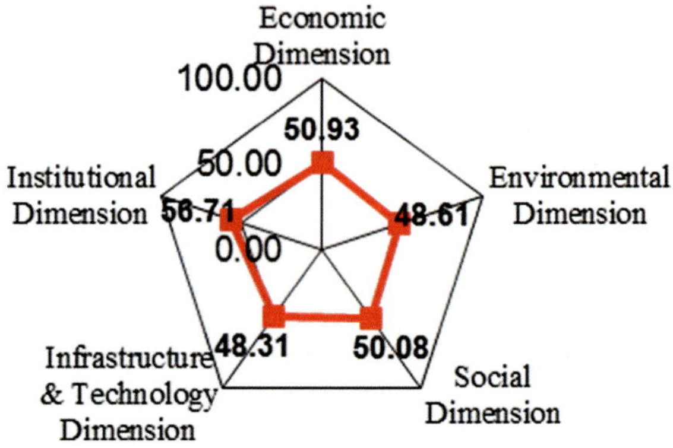

Fig. 3.7 The value of five dimensions

show one attribute as most sensitive, which is the proportion of off-farm activities managed by the community as shown in Fig. 3.6.

The proportion of the off-farm business sector that is managed and owned by the local community is very low. However, the average value of the score for on farm managed or owned by the local community is quite high. This shows that the existing subsystem on farm and off farm does not run simultaneously and synergistically. It is closely related to the institutional participation of farmers. The percentage of the farmers who are members of farmer groups and extension institutions are closely interrelated attributes. Compared with the number of farmers, the number of farmers who are members of farmer groups is much smaller. Institutional participation of farmers needs to be improved, because they are the main actors in the institutional development of the agropolitan area. Institutional participation of farmers who are dynamic and open to new innovations should serve as a dominant institution in the agropolitan area (Suwandi 2006).

Rap-Agrotropika ordinated analysis of the 62 attributes that affect the entire dimension shows the results of the multidimensional sustainability index value to be 50.30. These values are at an interval from 50.01 to 75.00 with a status quite sustainable. The value of the five dimensions is illustrated in a kite diagram as shown in Fig. 3.7.

Results of the analysis showed that to improve the status of agropolitan area sustainability, the dimensions of environmental, economic, and social have a number of important dimensions, without neglecting the role of other dimensions of technology and infrastructure as well as institutional. The analysis also confirms that the sustainability index value is actually on the status less sustainable, because the value of the multidimensional Rap-Agrotropika analysis is 50.30 with a status less sustainable. Rap-Agrotropika analysis can be applied to other locations, or used for other objectives, or at other times with certain conditions and attributes adjusted to local conditions.

In the development of an agricultural area, either existing areas or new areas, it is necessary to note that the characteristics include biophysical, economic, and socio-cultural aspects. The agropolitan area of Selupu Rejang belongs to the class of developed areas, because it has the characteristics of on-farm activities already developed, agricultural service-related institutions have started to be formed, and facilities and infrastructure are more complete, but for the development of downstream industry, activities are needed, and this is a field for further cultivation.

Development of a region cannot be separated from the socioeconomic conditions. Limited socioeconomic conditions, characterized by low levels of education and income, will affect the management of existing natural resources. Low level of knowledge about the environment will affect the way farmers implement agricultural cultivation.

The development of agropolitan areas is also associated with accessibility to economic resources and social services such as physical facilities, health, and basic needs. Accessibility to health services in Kecamatan Selupu Rejang is quite good, but based on the result of the interviews with respondents, people still feel difficulty to access to health services if their income is not sufficient. This shows that in the development of a region, not only the development of the cultivation system should be considered but also the fulfillment of the basic needs of the community. This has not yet received sufficient attention, because the program managers have stuck to the understanding that the development of the area is the development of cultivation systems and agribusiness.

Sustainability of the management of a region should be done with attention to the agroecosystem. Management of an agropolitan area should pay attention to the quality and suitability of the land. The production and productivity can be optimal and environmentally friendly. Area management should be in accordance with Indonesia's agroecosystem, namely, the tropical rainforest ecosystem. The cultivation of monoculture and intensive farming systems has caused a decline in the diversity that characterizes the tropical rainforest ecosystem in Indonesia. This causes the ecosystem to become unstable. Farming cultivation is done with a high input of fertilizers and pesticides. Implementation of environmentally friendly cultivation technology and proper conservation of land in ways that are socially acceptable to the community will be able to improve the sustainability of the region. Similarly, the type of crops cultivated should be in accordance with the agroecological zone mapping.

In relation to efforts to increase the value of diversity, environmentally friendly land management is needed, which is a way that can manage the trade-off between production (economy) and environmental quality. Tree-based planting pattern (agroforest) is one of the systems that can be a source of income for farmers and at the same time has the potential to maintain environmental services provided by forests, including biodiversity of forest origin (Noordwijk and Hairiah 2006b). Traditional farming systems such as yards can also maintain high ecosystem diversity especially in rural areas (Lahoti et al. 2013). This is not easy to achieve, due to the different interests and complexity of the problems in the field.

The development of the region for sustainability also cannot be separated from the supporting areas. Processing industry and agribusiness sub-terminals cannot develop well without the support of hinterland areas. The concept of sustainable agropolitan area management in the future will be built on factors or attributes that can provide most leverage and priorities. This concept is built on the understanding that the development of a region is not limited by administrative boundaries. The concept of agropolitan area management Selupu Rejang cannot stand alone but must be integrated with the hinterland area. Based on the master plan, there are four hinterland areas, i.e., Bermani Ulu subdistrict, Sindang Kelingi subdistrict, Curup subdistrict, and Ujan Mas subdistrict in Kepahiang Regency. It also shows that the future development of the agropolitan region requires coordination with other districts.

Management of agropolitan areas integrated with hinterland areas can provide transformation and supply of products, infrastructure, technology, markets, human resources, and technology. Hinterland district plays a role to supply the product. Sindang Kelingi and Curup subdistricts also play a role in improving human resources through counseling and strengthening farmer institutions related to the Agricultural Extension Institute (BPP) and human resources extension in both districts. Strengthening of the institution of farmers is important, because the development of agropolitan is essentially the process of creating added value in rural areas by the community itself. Because the strengthening and empowerment of the community becomes very important to note.

3.4 Conclusion

The status of multidimensional sustainability index is less sustainable. There are two dimensions are less sustainable, namely environmental dimension (48.61) and infrastructure and technology dimension (48.31). Dimensions of economic, social, and institutional sustainability status are quite sustainable with respective index values of 50.93, 50.08, and 56.71.

There are eight attributes as leverage factors. Leverage attributes of each dimension, namely: pesticides (ecological dimension); the proportion of agricultural employment, development of farm level price, and vegetable market (economic dimension); the level of knowledge about the environment (social dimension); rural community access to the centers of health and education services and the number of postharvest processing infrastructure (infrastructure and technology dimension); and the proportion of off farm which managed or owned by local communities (institutional dimension). To improve the sustainability status of the area, managers should pay attention to these leverage attributes.

References

Agropolitan Working Group of Rejang Lebong Regency (2007) Development of an independent agropolitan area. Paper presented at the National Agropolitan Session in Magelang, Central Java

Binder CR, Feola G, Steinberger JK (2010) Considering the normative, systemic and procedural dimensions in indicator-based sustainability assessments in agriculture. Environmental Impact Aseessment Review 30:71–81

Bosh RR, Diez-Unquera B, Ruiz R, Villalba D, Molina E, Joy M, Olaizola A, Bernues A (2012) An integrated sustainability assessment of Mediterranean sheep farms with different degrees of intensification. Agricultural Sistems 105:45–56

Dale VH, Beyeler SC (2001) Challenges in the development and use of ecological indicators. Ecol Indic 1:3–10

Environment Agency of Bengkulu Province (2013) Ulu Musi water quality monitoring report in Rejang Lebong and Kepahiang districts, Bengkulu Province

Environmental Research Center of Bengkulu University (2006) Preparation of action plan of natural resources and environment of Bengkulu Province

Friedmann J (1981) The active community: toward a political-territorial framework for rural development in Asia. Econ Dev Cult Chang 29:235–261

Lahoti S, Saito O, Mahalingam a, Gunatillete N, Irhamc VTH, Hitinayake G, Takeuchi K, Herath S (2013) Assessment of ecosystem services in homegardens systems in Indonesia, Sri Lanka, and Vietnam. Ecosyst Serv 5:124–136

Li M (2009) The choice of sustainable development model of China's. Agriculture Asian Social Science 5:91–93

Ministry of Agriculture (2002) Guidelines for the development of pilot programs agropolitan, Government of Indonesia

Muktamar Z, Sudjatmiko S, Toha B, Asteria M (2012) Farmers are sacrificing their health for production of vegetables. In: International conference on sustainable agriculture and food security: challenges and opportunities

Noordwijk MV, Hairiah K (2006a) Intensification of agriculture, soil biodiversity and agroecosystem function. Agri 28:3

Noordwijk MV, Hairiah K (2006b) Agricultural intensification, soil biodiversity, and agro-ecosystem function. Agri 28:3

Rao NH, Rogers PP (2006) Assessment of agricultural sustainability. Curr Sci 91:439–448

Rodrigues GS, Rodrigues IA, Buscjinelli CDA, Barros ID (2010) Integrated farm sustainability assessment for the environmental management of rural activities. Environ Impact Assess Rev 30:229–239

Rustiadi E (2007) Agropolitan: building a rural economy. Crestpent Press, Bogor

Sumantri S, Sukiyono K (2011) Perception and behavior land conservation in a variety of slope and its impact on the production of vegetable farming: a case study in Rejang Lebong the province of Bengkulu. Bumi Lestari Journal 11:138–146

Suwandi (2006) Strengthening rural economies institutional in agropolitan. In: Rustiadi E, Hadi S, Widhyanto (eds) Agropolitan: the concept of balanced rural-urban development. Crestpent Press, Bogor, pp 1–31

The Agropolitan Infrastructure and Facilities Support Unit (2005) Executive summary, bantek spatial plan of agropolitan area according to Law no. 26 Year 2007. Department of Public Works, Indonesia

Zhen L, Routray JK (2003) Operational indicators for measuring agricultural sustainability in developing countries. Environ Manag 1:34–46

Part II
Environmental Control

Chapter 4
The Diversity of Soil Fungus in and Around Termite Mounds of *Globitermes sulphureus* (Haviland) (Blattodea: Termitidae) and Response of Subterranean Termite to Fungi

Ikhsan Guswenrivo, Hideyuki Nagao, and Chow Yang Lee

Abstract Many researchers are trying to develop fungus-based biological control methods against insect pests, including termites. This study explored the termite–fungus relationship using *Globitermes sulphureus* (Haviland) as a model species. Fungal species in termite mounds of *G. sulphureus* were isolated, purified, and identified. These fungal species were then introduced to termites, and their interactions were characterized. The preliminary study found 24 species of fungus from 10 locations in and around the *G. sulphureus* mound, with the most common being 5 species belonging to *Trichoderma* sp., *Aspergillus* spp., and *Penicillium* spp. We found that termites practice a symbiont relationship with the five species of soil fungi with which they were experimented.

4.1 Introduction

Termites are related to the cockroach family and are regarded as pests. Since termite food is mainly wood and woody tissue, this insect causes many problems for wood users. Most termite species feed on dead plant materials above, at, or

I. Guswenrivo (✉)
Research Center for Biomaterials, Indonesian Institute of Sciences, Jl. Raya Bogor KM.46, Bogor 16911, Indonesia

Lab. Innovative Humano-habitability, RISH, Kyoto University, Gokasho Uji, Kyoto 611-0011, Japan
e-mail: ikhsan.guswenrivo@biomaterial.lipi.go.id

H. Nagao • C.Y. Lee
University Sains Malaysia (USM), Gelugor, P. Penang, 11800, Malaysia
e-mail: nagaoh@usm.my; chowyang@usm.my

B. McLellan (ed.), *Sustainable Future for Human Security*,
https://doi.org/10.1007/978-981-10-5430-3_4

under the soil surface and perform vital ecological roles in litter breakdown and in the recycling of mineral elements (Lavelle and Spain 2001), including dead foliage of grasses and other types of vegetation, woody materials such as roots and seeds, feces of higher animals, and other materials. As social insects, termites build nests to protect the entire colony from extreme climates and to defend against predators. In some species of termite, the nest functions as a place to store food (Noirot 1959; Thorne et al. 1996). Noirot and Johana (2000) identified three types of termite nest: subterranean (below ground level), epigeal (above the soil), and arboreal (on a trunk or tree branch and connected to the soil by crating mud tubes).

Globitermes sulphureus (Haviland) is commonly found in Malaysia, Singapore, and Vietnam. It has been classified under the family Termitidae. This species, which is highly prevalent in Malaysia, is easily identified based on the bright-yellow abdomens of the soldiers (Lee et al. 2003). It is a mound-building termite and builds epigeal nests. Noirot (1959) has categorized the mound of *G. sulphureus* as five layers: a royal chamber, pressed wood areas, the internal zone of the wall, the external layer of the wall, and the outer portion (bark). *G. Sulphureus* forage in the sand, feed on dead wood, and sometimes attack living trees and buildings (Roonwal 1970).

However, *G. sulphureus* is also beneficial to the environment as it breaks down wood debris and returns it to the soil. Besides termites, other microorganisms, such as fungi, also play important roles in decomposing dead wood and grasses. There are three functional groups of fungi: decomposers, mutualists, and pathogens. Around 100,000 out of 1.5 million estimated fungal species have been described by taxonomists (Hawksworth et al. 1995; Hawksworth 1991). Microbial populations in the soil are very diverse. Torsvik et al. (1996) calculated the presence of approximately 6000 bacterial genomes per gram of soil by taking the genome size of *Escherichia coli* (Migula) Castellani and Chalmers as a unit.

Being subterranean, *G. sulphureus* is exposed to various fungi and other pathogens that are abundant in the soil. Soil is a complex ecosystem where microorganisms live in heterogeneous communities, whereas the behaviors of individual species are often unknown due to the lack of suitable detection and identification techniques (Akkermans et al. 1994). Fungi play important physiological, mediative, and ecological roles in this ecosystem (Miller 1995). These roles include the decomposition of woody organic matter, the increase in nutrient uptake, the enhancement of plant resilience, and the improvement of soil structure (Jenkins 2005). In addition, some soil fungi have different functions and roles, and some are even harmful to other organisms, including termites.

Many studies have examined the interactions between termites and soil fungi (Sands 1970; Wong and Cheok 2001; Rosengaus et al. 2003). These relationships can be either symbiotic, attractant, antagonistic, pathogenic, or parasitic (Wong and Cheok 2001). More than 50 species of brown-rot fungi interact with subterranean termites, and these fungi can modify the nutrition, nest construction, survival, and caste development of termites (Amburgey and Beal 1977). The present research was carried out to investigate the number of fungal species and their populations from the inner, outer, and surrounding areas of mounds of *G. sulphureus* (Haviland)

and to focus on the interactions and relationships between subterranean *G. sulphureus* and soil fungi that were isolated from the termite nest.

4.2 Materials and Methods

4.2.1 Sampling of Nest Material

G. sulphureus nest materials were sampled from ten mounds at different locations on Penang Island, Malaysia. The height and width of the base of each mound were measured and recorded. Samples were taken from the inner and outer parts of the mound, and three samples (labeled A, B, and C) were taken from each section. The inner materials, taken from the second layer of the nest, 25–30 cm from the external wall, were dark brown and located near the royal chamber. The outer materials were taken from the fourth layer of the nest (internal wall), 5–7 cm from the external wall, and were dark and hard. Adjacent soil was sampled at 10–20 cm from each mound; three sections around the circumference of the mound were taken at depths of 5–15 cm. All the samples were kept in sterile paper bags and were then crushed with mortar porcelain until the maximum size of the material was 2 mm. All procedures were carried out under sterile conditions.

4.2.2 Isolation, Purification, and Identification of Fungi

Dilution plating was used to isolate fungi from the *G. sulphureus* carton. Five grams of *G. sulphureus* carton (inner and outer) and adjacent soil around the nest was mixed in 45 mL of sterile distilled water, shaken for 20 min, and kept under room conditions for 1 h to make a homogeneous suspension and to scrape the spores from the samples. Next, 10^{-2} and 10^{-3} dilutions were made by diluting 1 mL of suspension in 9 mL of sterile distilled water. Fifty microliters of suspension from each dilution was spread evenly over Rose Bengal Agar (RBA) surface with a sterile bent glass rod. The rod was sterilized by flaming with 95% ethanol before use. The inoculated dishes were incubated for 5 days at room temperature (25 ± 2 °C), after which the fungal colonies were counted. Each species of fungus was isolated and transferred into new PDA medium. Single-spore isolation was used to obtain pure fungal isolates. A piece of fungus was transferred to and diluted in sterile water and shaken. Fifty microliters of suspension was transferred on WA medium and incubated for 3 days, after which the single pure colony was removed and transferred on PDA medium to a petri dish. Morphological identification was made according to Moore-Landecker (1996). Microscopic observations were made by measuring the size of the hyphae and spores of the fungi under a compound microscope. The morphological identification was done up to the genus level.

Table 4.1 Observation of termite activity and fungal growth scores

Score	Termites	Fungi
1	90–100% dead	Fungi have not grown
2	Moribund, near death, still moving their antennae	Fungi growing, covering 10–20% of the bottom of the substrate
3	Not active, still moving slowly, grooming each other and themselves; 50–70% of termites dead	Fungi growing, covering 20–50% of the bottom of the substrate
4	Active, moving around the nest, grooming each other and themselves, collecting food; 20–50% of termites dead	Fungi growing, covering all of the bottom of the substrate and half of the surface area; 50–80% of the substrate overall
5	Active (healthy), moving around the nest, grooming each other and themselves, collecting food; <20% of termites dead	Fungi growing, covering 80–100% of the substrate overall, including all of the bottom and almost all of the surface

4.2.3 Response of G. sulphureus *to Fungi*

The impact of fungal concentration was a factor in this test. The experiment was conducted inside a glass jar (80 mm diameter x 150 mm height), termite carton (<20 mesh) as medium, and 40% water content. Five common fungal species isolated from the nest material were used in this test. These species were cultured in the lab on PDA medium. Two-hundred workers and 20 soldiers were used for each test. The activities of the termites and fungi were observed every 2 days up to 14 days. The observations of termite activity and fungal growth were done by giving a score for each level of termite activity and of fungal germination, respectively. The score ranges were 1–5 (see Table 4.1).

4.2.4 Statistical Analysis

All analyses were performed using SPSS version 16.0 software. Differences were tested by analysis of variance (ANOVA) and subsequent post hoc comparisons (Tukey HSD test). All tests were performed at a 95% confidence interval.

4.3 Results and Discussion

Detecting exactly which fungi were present in a soil sample was not easy, primarily because of the fastidious nature of the great majority of termite species. Hawksworth (1991) reported that there are 1.5 million species of fungi in soil and that only 10% of them have been described. Soil fungi are a group of microorganisms that can actively grow in close association with other organisms or can exist in a dormant form. Saprobic fungi are the largest group of fungal species that occur in

Table 4.2 Colony and species numbers from direct and dilution plating methods from *G. sulphureus* (Haviland) mound carton and adjacent soil

Sampled part of the mound	Dilution plating method		Direct plating method	
	Species number	CFU/gram soil (×1000)	Species number	CFU/gram soil (×1000)
Inner part	4.20 ± 0.92	46.90 ± 26.66	3.00 ± 0.95	–
Outer part	6.60 ± 1.07	31.51 ± 2.32	4.30 ± 0.95	–
Adjacent soil	9.50 ± 1.08	9.29 ± 1.55	4.29 ± 1.03	–

soil, and they have an important role in the decomposition of cellulose, hemicellulose, and lignin compounds, which in turn contribute to the carbon cycle (Mahmood et al. 2006; Saravanakumar and Kaviyarasan 2000). In addition, these catabolic activities enable fungi to grow in innutritious substrates. A lot of studies, using various methods, have described the diversity of soil fungi (Viaud et al. 2000; Bridge and Spooner 2001; Hawksworth 2001; Celine et al. 2004). Yet, few studies have described the diversity of soil fungi and the interaction between fungi and termites. This report discusses the diversity of the fungal community in *G. sulphureus* mounds as determined by isolating, culturing, and counting colonies using the dilution plating method.

Fungal species composition from the termite carton and adjacent soil can be analyzed by either the direct or the dilution plating method. Our preliminary study showed that the number of fungal species from the dilution plating method was more diverse compared to the direct plating method (Table 4.2). The fungal species grown by direct plating were also found by the dilution plating method, so the latter was used to isolate fungal species in *G. sulphureus* mounds.

Dilution plating is a method to isolate a fungus by immersing a sample in sterile water and culturing the fungus. Here, dilution plating showed that the inner part of a mound contains fewer species with high colony numbers. As for the outer part, this method revealed high numbers of fungal species but with low colony numbers. The different numbers of fungal species from the inner and outer parts could be related to the moisture content of the soil. It is supported by the data of *G. sulphureus* nest, which are the moisture content of the *G. sulphureus* nest at the inner part of the mound was higher than that at the outer part or in the adjacent soil24.

Based on the results shown in Table 4.2, the isolation of fungal species from *G. sulphureus* mounds was carried out using the dilution plating method, and RRBA medium was used to grow the fungi. Figure 4.1 shows the numbers of colonies and species of soil fungi from each sampling spot (inner, outer, and adjacent). The inner part of the mound was less diverse than the outer or the adjacent soil. In previous studies, the species number suggested that the inner, outer, and adjacent samples had least to most diversity, in that order. The mean species number from the inner part was less than that of the outer part or that of the adjacent soil: 3.55 ± 0.21, 5.82 ± 0.38, and 7.08 ± 0.45 species, respectively. On the other hand, the inner part had more colonies than either the outer or adjacent part: 4.77 ± 0.56, 2.44 ± 0.14, and 1.19 ± 0.31, respectively.

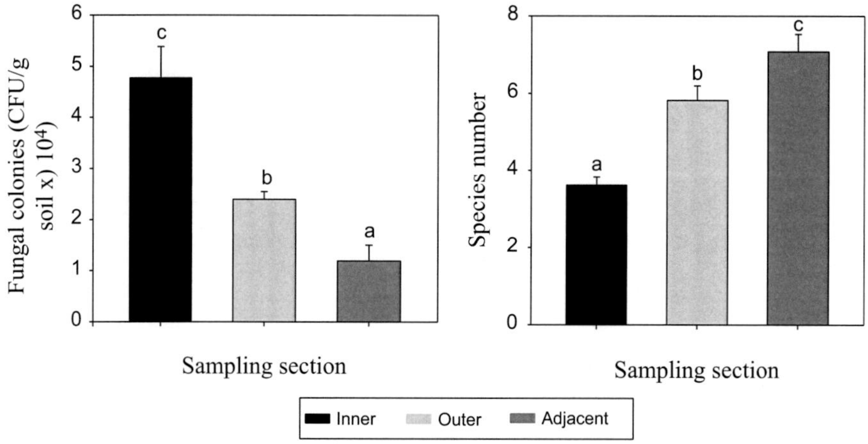

Fig. 4.1 Fungal colony and species numbers from three-sample section (inner, outer, and adjacent soil) of *G. sulphureus* mound

The growth of fungal species was correlated with the nutrient content in *G. sulphureus* mounds, each part of which had a different nutrient composition. The inner part contained more cellulose and nitrogen compared to the outer part (Guswenrivo et al. 2011). Griffin (1972) reported that carbon, nitrogen, and minor nutrients (such as phosphorus and sulfur) are crucial for fungal growth, as is the ratio between carbon and nitrogen. Fungal growth is known to be affected by other factors, as well, such as moisture, temperature, the quality of resources, competition, and local processes such as predation and other biota (Celine et al. 2004). The *G. sulphureus* nest is built from soil together with their saliva and also from undigested food that is rich in nutrients such as nitrogen and carbon (Krishna and Weesner 1970; Brauman 2000). In other words, the inner part of the mound possesses optimal conditions (humidity and acidity) for fungal growth.

All of those conditions should support fungal growth, but the present results turn out to be different. The inner part of the mound, which is highly suitable for fungal growth, lacked fungal populations when compared to the outer part or even the adjacent soil. This was because the termites concentrated in the center of the nest. Other factors have also been found to affect and inhibit fungal growth.

The number of termites present in a nest affects fungal availability, since *G. sulphureus* grooms in the nest. Grooming is a social behavior; termites clean their bodies, the nest, and their nestmates. Termite saliva reportedly contains an antifungal component that can inhibit fungal growth. In light of this, some fungal species were not expected to survive or might exist in a dormant stage. Another factor is the competition for food sources between *G. sulphureus* and fungi. The center part of the nest was full of termites (workers, soldiers, and nymphs), so fungal growth in the mound would be restricted. Overall, 11 species of fungi were isolated in the inner part, 15 species in the outer part, and 24 in the adjacent soil.

Table 4.3 Diversity of fungal species isolated from inner part, outer part, and adjacent soil of *G. sulphureus* (Haviland)

No.	Sample area	Group	Genus
1.	Inner part	*Ascomycota*	*Aspergillus* sp. (two species)
			Penicillium sp. (four species)
			Trichoderma sp. (one species)
		Zygomycota	*Mucor* sp. (one species)
		Unidentified fungi (three species)	
2.	Outer part	Ascomycota	*Aspergillus* sp. (two species)
			Penicillium sp. (six species)
			Trichoderma sp. (one species)
			Fusarium sp. (one species)
		Zygomycota	*Mucor* sp. (one species)
		Unidentified fungi (four species)	
3.	Adjacent soil	*Ascomycetes*	*Aspergillus* sp. (two species)
			Penicillium sp. (nine species)
			Trichoderma sp. (one species)
			Fusarium sp. (one species)
		Zygomycota	*Mucor* sp. (one species)
		Unidentified fungi (ten species)	

The five common fungal species isolated from the nest material and adjacent soil included one species of *Trichoderma*, two of *Aspergillus*, and two of *Penicillium*.

Table 4.3 shows that most fungal species in the inner part were from groups of Ascomycetes. Fungi such as *Trichoderma* sp., *Aspergillus* sp., and *Penicillium* sp. were the dominant species of fungi isolated from all three locations. Being subterranean, *G. sulphureus* colonies are exposed to various fungi and other pathogens that can be found in the soil. In interactions between fungi and termites, fungi might function as symbionts, attractants, antagonists, pathogens, or even parasites to the termites (Wong and Cheok 2001). This was confirmed by Corinne (2000), who stated that the presence of fungi in a termite nest can include compounds that attract or repel termites and can also affect termite nutrition. *Gloeophyllum trabeum* and *Serpula lacrymans* each contain a substance attractive to termites (Esenther et al. 1961; Matsumura et al. 1969; Sugamoto et al. 1990). Smythe et al. (1970) reported that *Reticulitermes flavipes* survived better on decayed wood than on sound wood. This was due to some fungal species having modified fatty acids, affecting their protein and amino acid compositions or even serving as a food source for *R. flavipes* (Carter et al. 1972; Carter and Smythe 1973; Waller et al. 1987; Waller 1993).

Subterranean termites live in soil, together with more than one million species of fungi. Termites live in colonies with high-density populations and humid conditions, making them susceptible to fungal infection (Vargo et al. 2003; Yanagawa and Shimizu 2007). However, there is a lack of information on interactions between termites and soil fungi. Most information concerns interactions and relationships

Table 4.4 Termite mortality and food consumption (%) during exposure to soil fungi at four different incubation times after inoculation (in weeks)

Observation factor	Age	Fungal species *Trichoderma* sp.	*Aspergillus* sp. (B)	*Aspergillus* sp. (G)	*Penicillium* sp. (A)	*Penicillium* sp. (B)	Control
Termite mortality by week (%)[1]	0	50.4 ± 42.47 a	48.5 ± 14.22 a	68.4 ± 11.12 a	78.9 ± 14.15 a	80.89 ± 13.89 a	26.5 ± 14.88
	1	100.0± 0 b	100.0± 0 b	100.0± 0 b	100.0± 0 b	100.0± 0 b	
	2	100.0± 0 b	100.0± 0 b	100.0± 0 b	55.7 ± 12.52 b	77.9 ± 17.55 b	
	4	100.0± 0 b	100.0± 0 b	100.0± 0 b	100.0± 0 b	100.0± 0 b	
Food consumption by week (%)[1]	0	2.42 ± 0.60 b	2.38 ± 0.43 b	2.47 ± 0.27 b	2.28 ± 0.32 c	2.21 ± 0.43 c	6.20 ± 0.76
	1	0.63 ± 0.22 a	0.54 ± 0.18 a	0.75 ± 0.20 a	0.76 ± 0.12 a	0.71 ± 0.22 a	
	2	0.68 ± 0.20 a	0.77 ± 0.20 a	0.66 ± 0.35 a	1.18 ± 0.35 b	1.24 ± 0.47 b	
	4	0.70 ± 0.11 a	0.57 ± 0.16 a	0.57 ± 0.57 a	0.48 ± 0.06 a	0.65 ± 0.65 a	

[1]Means followed by the same letter within the same row are not significantly different ($P > 0.05$; Tukey HSD), $N = 5$

between Macrotermitinae and a fungus-growing termites subfamily of Termitinae (Aanen et al. 2007), and other Termitinae subfamilies, and also on the symbiosis between termites and wood decay (Cornelius et al. 2002). An experiment on the interaction between *G. sulphureus* and soil fungi was conducted for a period of 14 days. Termite activities such as grooming behavior, movement, mortality, and food consumption were recorded. Table 4.4 shows the mortality and food consumption activities of *G. sulphureus* after the termites were exposed to five species of soil fungi of four different fungal growth ages after the fungi were inoculated after that period.

Apparently, the age of a soil fungus affects termite mortality. Their ability to survive and their resilience on substrates containing fungi were tested with five different species of soil fungi in four different age groups (0, 1, 2, and 4 weeks). Termites practice hygiene and grooming activities to protect themselves and the colony from contamination. Beginning on the first day of a fungal incubation period, the fungus grew very productively, with both an increasing volume of individual cells, increasing overall numbers of cells, and growing biomass (Reeslev and Kjøller 1995). During the second week of incubation, fungi would start to secrete secondary metabolites (enzymes) that may be useful, harmful, or toxic to other organisms (Maheshwari et al. 2000). Each fungus secreted enzymes at different rates, e.g., *Phanerochaete* starts to increase its enzyme secretion after 8 days of incubation (Johnson et al. 1994), and *Colletotrichum gloeosporioides* started to secrete enzymes after 4 days of incubation (Onofre et al. 2011). Fungi began to decrease their production of secondary metabolites when they reached 4 weeks of incubation.

Termite mortality on growth day 0 of each fungal species differed significantly ($P < 0.05$) depending on the fungal incubation period, i.e., after 1, 2, and 4 weeks. This was supported by the data on the termites' consumption of a wood block, which differed significantly ($P < 0.05$) among the 1-, 2-, and 4-week fungal growth periods; at fungal growth day 0, *G. sulphureus* consumed more wood than at the other time points (Table 4.4). The observation of termite interactions showed that the survival rate of *G. sulphureus* decreased with increasing time. Termite mortality increased over time, reaching 100% at the 14-day observation after being exposed to fungi at 1, 2, and 4 weeks of growth, whereas for termites exposed to fungi on growth day 0, mortality of termite at the 14-day observation ranged between 48% and 80%. These differences in termite mortality depending on the length of time after incubation were caused by the compounds released by the fungi that influenced the interactions between them and the termites. These compounds are commonly associated with the sporulation process in microorganisms, including fungi, and are often called secondary metabolites (Bu'Lock 1961; Sekiguchi and Gaucher 1977). Fungi usually begin producing these compounds upon entering a stationary or resting phase (Bu'Lock 1961).

On the first day of observation, there were high numbers of healthy termites, but the number gradually decreased up to day 14. This trend was demonstrated by the activities of the termites. The activity levels and healthiness of the termites decreased with time (Fig. 4.2), as reflected in the decreasing numbers of healthy

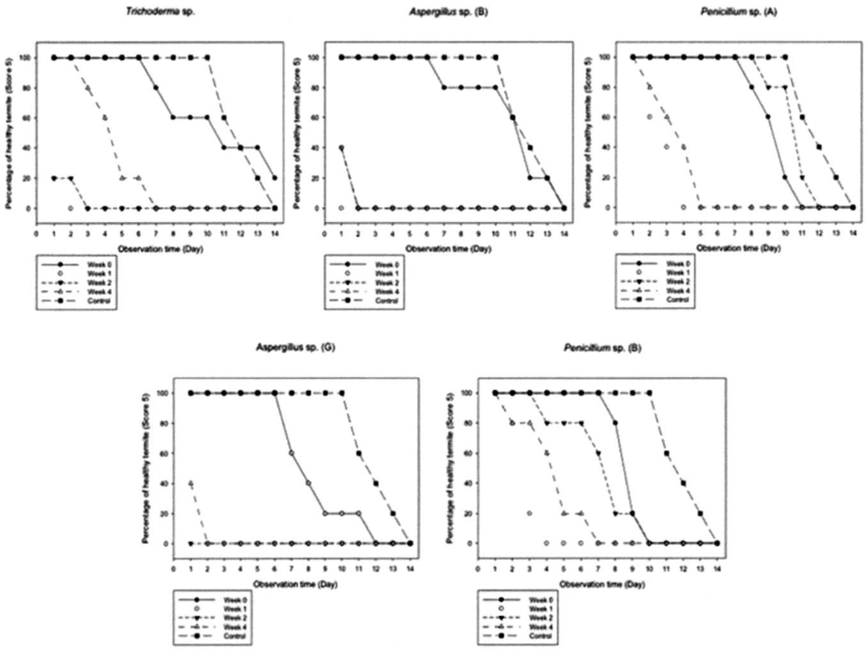

Fig. 4.2 The activity of *G. sulphureus* Haviland upon exposure to five different species of fungi (*Trichoderma* sp., *Aspergillus* sp. (A), *Aspergillus* sp. (B), *Penicillium* sp. (A), *Penicillium* sp. (B)) at different incubation periods after inoculation based on the percentage of healthy termites (score 5)

termites that still survived. The movement of termites and their grooming activities began to decline. Fungal growth increased with time, as indicated by the presence of mycelia and hyphae. Figure 4.2 shows the stage of fungal growth when the fungi came into contact with *G. sulphureus*. Due to the different lengths of time after incubation, there were different stages of fungal growth when the experiment was started. The 1-, 2-, and 4-week fungi completely covered the surface of the substrate for.

Among the factors that would influence the interactions between termites and soil fungi is the population density of fungi in the soil. In this chapter, two different population densities were tested, and termite mortality and food consumption were observed (Table 4.5). There was a significant difference ($P < 0.05$) between the two concentrations; 10^5 CFU of soil fungi was a concentration that *G. sulphureus* could tolerate, with the mortality rate varying according to the fungal species. However, at 10^7 CFU of soil fungi, termite mortality was 100% regardless of the fungal species (Table 4.5). On the other hand, food consumption did not differ significantly ($P > 0.05$) between the two concentrations (10^5 CFU and 10^7 CFU; Table 4.5).

The inner part of the *G. sulphureus* (Haviland) mound contains lower concentrations of soil fungi compared to the outer part and adjacent soil. This corroborated well a report by Chouvenc and Su (2010), where subterranean termites exhibited an

Table 4.5 Termite mortality (%) and food consumption after exposure to two concentrations (CFU) of soil fungi for 14 days

No.	Fungal species	N	Termite mortality (%)		Food consumption (%)	
			10^5 CFU	10^7 CFU	10^5 CFU	10^7 CFU
1	*Trichoderma* sp.	5	48.7 ± 35.28 a[1]	100.0± 0 b	2.61 ± 0.74 a[1]	1.20 ± 0.31 a
2	*Aspergillus* sp. (G)	5	94.5 ± 10.65 a	100.0± 0 b	1.50 ± 0.46 a	0.70 ± 0.05 a
3	*Aspergillus* sp. (B)	5	98.5 ± 2.12 a	100.0± 0 b	1.72 ± 0.23 a	0.64 ± 0.12 a
4	*Penicillium* sp. (A)	5	73.5 ± 11.07 a	100.0± 0 b	2.57 ± 0.32 a	0.79 ± 0.23 a
5	*Penicillium* sp. (B)	5	95.9 ± 3.66 a	100.0± 0 b	2.08 ± 0.39 a	0.78 ± 0.74 a
6	Control	5	27.5 ± 14.45		5.84 ± 0.81	

[1]Means followed by the same letter within the same row are not significantly different ($P > 0.05$; Tukey HSD)

ability to protect and defend themselves and their entire colony from fungi by three major defense mechanisms: grooming, cellular encapsulation, and gut antifungal activity. Those authors explained that the relationship between termites and entomopathogenic fungi is multilevel and that each level is a barrier against fungi that prevents fungi from thriving and completing their life cycle.

In another study, exposure to soil fungi elicited various behaviors from termites, including hygienic, attacking, cannibalistic, and burying behaviors (Noirot and Darlington Johana 2000). This observation showed that *G. sulphureus* workers removed fungal mycelia from the surrounding food sources. On the other hand, the observation also showed the termites groomed themselves as well as their nestmates. Yanagawa et al. (2009, 2010) found that the grooming activity of subterranean termite *C. formosanus* did not change when it came into contact with entomopathogenic fungi. Recently, Yanagawa (2011) reported that termite grooming activity differed significantly and increased when uninfected termites mixed with termites infected with entomopathogenic fungi. The termites exhibited this behavior from the very first day after being exposed to the infected termites, and termite was focused in areas where food was available. Termite mortality eventually increased with time. This also indicates that the termite survival was stronger at substrates contaminated with soil fungi.

Fungal growth can be inhibited by termites through hygienic and grooming activities (Fig. 4.3). This is because when termites groom, they use their saliva to cleanse themselves and their nestmates from fungal spores/mycelia that were attached to them. Lamberty et al. (2001) found that the saliva secreted by the higher termite *Pseudocanthotermes spiniger* during grooming contains the antifungal peptide termicin. The same was true of *Nasutitermes* sp., in whose saliva Bulmer and Crozier (2004) found a related antifungal peptide. This corroborated well the observation in this study that *G. sulphureus* inhibited the growth of soil fungi (fungi age 0 week) and cleaned the surface of the substrate from mycelia (fungal ages 1, 2, and 4 weeks). Fungi at age 0 week grew vines on the bottom of the substrate, while the substrate surface was not yet covered with fungal mycelia. With fungi at ages 1, 2, and 4 weeks, the substrate was fully covered with mycelia, but termites started

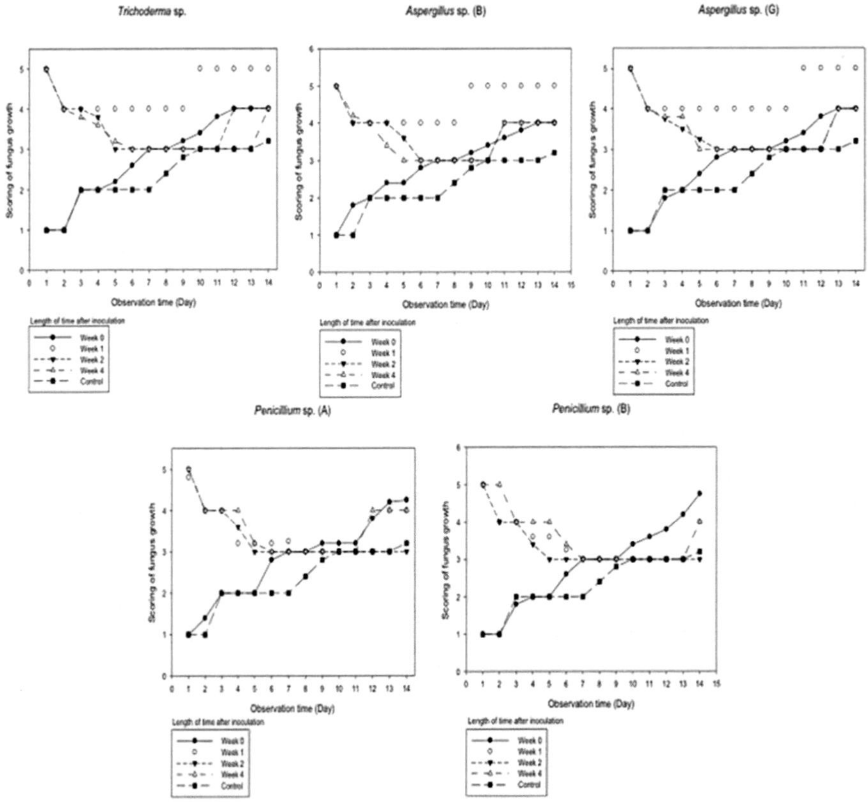

Fig. 4.3 Growth rates of fungi (*Trichoderma* sp., *Aspergillus* sp. (G), *Aspergillus* sp. (B), *Penicillium* sp. (A), and *Penicillium* sp. (B)) in the presence of *G. Sulphureus*

to clean up the substrate surface and moved the mycelia aside. After several days of exposure to fungi, *G. sulphureus* started to make tunnels under and around food sources. This indicated that *G. sulphureus* can protect themselves from contamination.

The antifungal activity was observed in this colony when *G. sulphureus* removed mycelia of soil fungi aside by picking them up with their mandibles and pushing them away from food sources. Some fungal spores will attach to the termite body, while others will enter the abdomen, which can kill a termite. Therefore, grooming activity, where nestmates clean each other with their glossae and excrete the spores after digesting them, plays an important role (Yanagawa and Shimizu 2005, 2007). The inhibition of fungal growth shown by the termite gut and/or termite body indicated that *G. sulphureus* produced and secreted antifungal compounds. Previous reports showed that the termite intestine has a fungistatic ability against hyphal growth and that the termite gut resists the growth of conidia once they pass through it (Bao and Yendol 1971; Kramm and West 1982; Boucias et al. 1996; Rosengaus et al. 1998; Siderhurst et al. 2005).

On the other hand, the concentration of fungal spores in soil also affects the activities and mortality of *G. sulphureus*. Previous studies showed that the presence of fungal spores at a particular concentration could increase the rate of termite mortality up to 100% within a few hours, whenever contact was possible between the termites and fungi (Yanagawa and Shimizu 2007). Yanagawa and Shimizu (2007) explained that a lower concentration of fungal conidia in substrate did not affect termite activity due to the termites' hygienic, grooming, and antifungal activities. *G. sulphureus* showed the same reactions when it was exposed to the lower of two concentrations of soil fungi (10^5 CFU) used in this study. Termite mortality was lower at 10^5 CFU than at 10^7 CFU. This shows that *G. sulphureus* was able to survive at this concentration of fungal spores through hygiene and grooming activity.

In fact, the inner part of the *G. sulphureus* mound has fewer species of soil fungi than the outer part of the mound or the adjacent soil. The reverse was true for the number of colonies: the inner part had more than the other parts. The Ascomycetes group, whose members include *Trichoderma* sp., *Aspergillus* sp., and *Penicillium* sp., among others, accounts for the majority of fungal species inside the mound. The different numbers of fungal species between the inner part, outer part, and adjacent soil of the *G. sulphureus* (Haviland) mound showed that the termites interact with soil fungi. The interactions of *G. sulphureus* with soil fungi have been tested with five species of soil fungi (*Trichoderma* sp., *Aspergillus* sp. (two species), and *Penicillium* sp. (two species)), revealing that the density of the fungi affected the termites' ability to stand up to and adapt to those fungi. In addition, the ability of *G. sulphureus* to survive and interact with soil fungi could be supported by the antifungal compounds produced in the termite gut.

References

Aanen DK, Vera IR, HdFL H, Mitchell J, ZdB W, Slippers B, Corinne RL, Boomsma JJ (2007) Patterns of interaction specificity of fungus-growing termite and termitomyces symbionts in South Africa. BMC Evol Biol 7:115

Akkermans ADL, Mirza MS, Harmsen HJM, Blok HJ, Sessitsch A, Akkermans WM (1994) Molecular ecology of microbes: a review of promises, pitfalls and true progress. FEMS Microbiol Rev 15:1850194

Amburgey TL, Beal RH (1977) White rot inhibits termite attack. Sociobiology 3:35–38

Bao LL, Yendol WG (1971) Infection of the eastern subterranean termite, *Reticulitermes flavipes* (Kollar) with the fungus *Beauveria bassiana* (Balsamo) Vuill. Entomophaga 16:343–352

Boucias DG, Stokes C, Storey G, Pendland JC (1996) Effect of imidacloprid on the termite, *Reticulitermes flavipes* and its interaction with insect pathogens. Pfanzenshutz-Natrichten Bayer 9:103–144

Brauman A (2000) Effect of gut transit and mound deposit on soil organic matter transformations in soils feeding termite: a review. Eur J Soil Biol 36:117–125

Bridge P, Spooner B (2001) Soil fungi: diversity and detection. Plant Soil 232:147–154

Bu'Lock JD (1961) Intermediary metabolism and antibiotic synthesis. Adv Appl Microbiol 3:293–342

Bulmern M, Crozier RH (2004) Duplication and diversifying selection among termite antifungal peptide. Mol Biol Evol 21:2256–2264
Carter FL, Smythe RV (1973) Effect of sound and *Lenzites* decayed wood on the amino acid composition of *Reticulitermes flavipes*. J Insect Physiol 19:1623–1629
Carter FL, Dinus LA, Smythe RV (1972) Effects of wood decayed by *Lenzites trabea* on the fatty acid composition of the eastern subterranean termite, *Reticulitermes flavipes*. J Insect Physiol 18:1387–1393
Celine RA, Brygoo Y, Harry M (2004) Ascomycete diversity in soil-feeding termite nests and soils from a tropical rainforest. Environ Microbiol 6(5):462–469
Chouvenc T, Su N-Y (2010) Apparent synergy among defense mechanisms in subterranean termites (Rhinotermitidae) against epizootic events – the limits and potential for biological control. J Econ Entomol 103:1327–1337
Corinne RL (2000) Chapter 14: Symbiosis with fungi. In: Abe T et al (eds) Termite: Evolution, sociality, symbioses, ecology. Kluwer Academic Publishers, Dordrecht, pp 289–306
Cornelius ML, Daigle DJ, Connick WJ, Tellez M, Williams KS, Lovisa MP (2002) Interaction between Formosan subterranean termites (isopteran: Rhinotermitidae) and wood decay fungi. Proceedings of the 4th international conference on urban pests. pp 319–324
Elizabeth Moore Landecker (1996) Fundamentals of the fungi, 4th edn. Prentice-Hall, Englewood Cliffs
Esenther GR, Allen TC, Casida JE, Shenefelt RD (1961) Termite attractant from fungus – infected wood. Science 134(3471):50
Griffin DM (1972) Ecology of soil fungi. Chapman and Hall, London
Guswenrivo I, Nagao H, Lee CY 2011 Analysis of cellulose and nitrogen content of nest materials of a higher termite *Globitermes sulphureus* (Haviland). In: The 8th Pacific Rim Termite Research Group Conference, Bangkok, Thailand. pp 83–86
Hawksworth DL (1991) The fungal dimension of biodiversity: magnitude, significance and conservation. Mycology Research 95:641–655
Hawksworth DL (2001) The magnitude of fungal diversity: the 1.5 million species estimate revisited. Mycol. Res 105(12):1422–1432
Hawksworth DL, Kirk PM, Sutton BC, Pegler DN (1995) Ainsworth and Bisby's dictionary of the fungi. CAB International, Wallingford, p 424
Jenkins A (2005) Soil Biology. In: Soil Biology Basic. State of New South Wales. Department of Primary Industries, Richmond
Johnson CR, Cullen D, Lamar RT (1994) Manganese peroxidases of the white rot fungus *Phanerochaete sordida*. Appl Environ Microbiol 60(2):599–605
Kramm KR, West DF. Termite pathogens: effects of ingested *Metarhizium*, *Beauveria* and *Gliocladium* conidia on worker termites (Reticulitermes sp.). J Invertebr Pathol. 1982; 40, 7–11
Krishna K, Weesner FM. Biology of termites. Academic Press, London. 1970; 346
Lamberty M, Zachary D, Lanot R, Bordereau C, Robert A, Hoffman JA, Bulet P (2001) Insect immunity constitutive expression of a cysteine-rich antifungal and a linear antibacterial bacterial peptide in a termite inset. J Biol Chem 276:4085–4092
Lavelle P, Spain AV. Soil ecology, Kluwer Academic Publishers, London. 2001; 294–316
Lee CY, Yap J, Ngee PS, Jaal Z (2003) Foraging colonies of higher mound-building subterranean termite, *Globitermes sulphureus* (Haviland) in Malaysia. Japanese Journal Environmental Entomology and Zoology 14:105–112
Maheshwari R, Bharadwaj G, Bhat MK (2000) Thermophilic fungi: their physiology and enzymes. Microbiol Mol Biol Rev 64(3):461–488
Mahmood K, Wei-jun Y, Nazir K, Zahid IR, Abdullah GA (2006) Study of cellulolytic soil fungi and two nova species and new medium. J Zhejiang Univ Sciences 7:459–466
Matsumura F, Tai A, Coppel HC (1969) Termite trail-following substance, isolation and purification from *Reticulitermes virginicus* and fungus-infected wood. J Econ Entomol 62:599–603
Miller SL (1995) Functional diversity in fungi. Can J Bot 73:S50–S57

Noirot C (1959) Le nid de *Globitermes sulphureus* (Haviland) au cambodge. Insect Soc 6:259–269

Noirot C, Darlington Johana PEC. Termite nests: architecture, regulation and defence. *In* T. Abe *et al.* Termite: Evolution, sociality, symbioses, ecology. Kluwer Academic Publishers, Dordrecht. 2000; 121–139

Onofre SB, Steilmann P, Bertolini J, Rotta D, Francini AS, Kagimura Y, Groff SA, Mazzali L (2011) Amylolytic enzymes produced by the fungus *Colletotrichum gloeosporioides* in rice semi-solid fermentation. J Yeast Fungal Res 2(3):28–32

Reeslev M, Kjøller A (1995) Comparison of biomass dry weight and radial growth gates of fungal colonies on media solidified with different gelling compounds. Appl Environ Microbiol 61:4236–4239

Roonwal ML (1970) Termites of the oriental region. In: Krishna K, Weesner FM (eds) The Biology of Termites, vol II. Academic Press, New York, pp 315–391

Rosengaus RB, Guldin MR, Traniello JFA (1998) Inhibitory effect of termite fecal pellets on fungal spore germination. J Chem Ecol 24:1697–1706

Rosengaus RB, Moustakas JE, Calleri DV, JFA T (2003) Nesting ecology and cuticular microbial loads in dampwood (*Zootermopsis angusticollis*) and drywood termites (*Incisitermes minor, I. schwarzi, Cryptotermes cavifrons*). J Insect Sci 3:31–37

Sands WA (1970) The association of termites and fungi. In: Krishna K, Weesner FM (eds) The Biology of Termites, vol I. Academic Press, New York, pp 495–542

Saravanakumar K, Kaviyarasan V (2000) Seasonal distribution of soil fungi and chemical properties of montane wet temperate forest types of Tamil Nad. African J Plant Sci 4:190–196

Sekiguchi J, Gaucher GM (1977) Conidiogenesis and secondary metabolism in *Penicillium urticae*. Appl Environ Microbiol 33:147–158

Siderhurst MS, James DM, Blunt TD, Bjostad LB (2005) Antimicrobial activity of norharmane against the entomopathogenic fungus Metarhizium anisopliae (Metsch.) and the caste and phylogenetic distribution of this defense in termites (Insecta: Isoptera). Sociobiology 46:563–577

Smythe RV, Carter FL, Bacter CC (1970) Influence of wood decay on feeding and survival of the eastern subterranean termite, *Reticulitermes flavipes* (Isoptera: Rhinotermitidae). Ann Entomol Soc Am 64:59–62

Sugamoto K, Minamide Y, Katsuda Y, Nishimoto K 1990 Abstract of 2nd annual meeting of the Japan society for Environmental Entomology and Zoology

Thorne BL, Collins MS, Bjorndal KA (1996) Architecture and nutrient analysis of arboreal carton nests of two Neotropical *Nasutitermes* species (Isoptera: Termitidae), with notes on embedded nodules. Fla Entomol 79:27–37

Torsvik V, Sorheim R, Goksoyr J (1996) Total bacterial diversity in soil and sediment communities–a review. J. Ind. Microbiol 17:170–178

Vargo EL, Husseneder C, Grace JK (2003) Colony and population genetic structure of the Formosan subterranean termite, *Coptotermes formosanus*, in Japan. Mol Ecol 12:2599–2608

Viaud M, Pasquier A, Brygoo Y (2000) Diversity of soil fungi studied by PCR-RFLP of ITS. Mycol Res 104(9):1027–1032

Waller DA (1993) Response by *Reticulitermes* (Isoptera: Rhinotermitidae) and *Cryptocercus punctulatus* (Blattaria:Cryptocercidae) to wood infected with the green-stain ascomycete, *Chlorociboria aeruginascens aeruginascens*. Sociobiology 23:103–108

Waller DA, La Fage JP, Gilbertson RL, Blackwell M (1987) Wood-decay fungi associated with subterranean termites (Rhinotermitidae) in Louisiana. Proc Entomol Soc Wash 89:417–424

Wong AHH, Cheok KS 2001 Observations of termite-fungus interactions of potential significance to wood biodeterioration and protection. Timber Technology Bulletin 24. Kuala Lumpur

Yanagawa A (2011) Fujiwara-Tsujii, N Akino, Yoshimura T, and Shimizu S. Musty odor of entomopathogens enhances disease-prevention behaviors in the termite *Coptotermes formosanus*. J Invertebr Pathol 108(1):1–6

Yanagawa A, Shimizu S (2005) Defense strategy of the termite, *Coptotermes formosanus* Shiraki to entomopathogenic fungi. Jpn J Envir Entomol Zool 16:17–22

Yanagawa A, Shimizu S (2007) Resistance of the termite, *Coptotermes formosanus* Shiraki to *Metarhizium anisopliae* due to grooming. BioControl 52(1):75–85

Yanagawa A, Yokohari F, Shimizu S (2009) The role of antennae in removing entomopathogenic fungi from cuticle of the termite, *Coptotermes formosanus*. J Insect Sci 9

Yanagawa A, Yokohari F, Shimizu S (2010) Influence of fungal odor on grooming behavior of the termite, *Coptotermes formosanus* Shiraki. J Insect Sci 10

Chapter 5
Screening of Strains and Media of *Termitomyces eurrhizus* Collected from Japan Toward Nutritional Analyses

Kazuko Ono, Tsuyoshi Yoshimura, and Kazuhiko Kinjo

Abstract Twenty-seven strains of *Termitomyces eurrhizus*, all from Okinawa Prefecture, Japan, were served for screening with eight artificial media. Four strains were selected for their growth, media specificity, and handling for future experiments on nutritional analyses. In addition, the matsutake medium was found to have a good affinity to these strains.

5.1 Introduction

Termitomyces is a member of the family Lyophyllaceae in the order Agaricales. The fungi in this genus are known to have a symbiotic relationship with Macrotermitinae termites and are called termite mushrooms. Their distribution range includes tropical or subtropical areas from Africa to Southeast Asia (Aanen and Boomsma 2007). For the termites, the fungal symbiont is thought to be the main decomposer of lignocellulose in dead plant materials which are collected by termites. However, the exact function of the symbiosis is still unclear and seems to differ among genera and species (Rouland-Lefèvre et al. 1991; Hyodo et al. 2003; Ohkuma 2003). Four hypotheses about the symbiosis are proposed at present (Nobre et al. 2011): (1) *Termitomyces* is an additional protein-rich food source; (2) *Termitomyces* has a role in lignin degradation, which facilitates the access to cellulose; (3) *Termitomyces* decreases the C/N ratio of foraged products by

K. Ono (✉) • T. Yoshimura
Research Institute for Sustainable Humanosphere (RISH), Kyoto University, Gokasho, Uji-shi, Kyoto 611-0011, Japan
e-mail: onokazu@rish.kyoto-u.ac.jp; tsuyoshi@rish.kyoto-u.ac.jp

K. Kinjo
Faculty of Agriculture, University of the Ryukyus, 1 Senbaru, Nishihara-cho, Nakagami-gun, Okinawa 903-0213, Japan
e-mail: ykimjo@agr.u-ryukyu.ac.jp

B. McLellan (ed.), *Sustainable Future for Human Security*,
https://doi.org/10.1007/978-981-10-5430-3_5

metabolizing carbohydrates; and (4) *Termitomyces* provisions cellulases and xylanases to work synergistically and/or complementarily with endogenous termites enzymes.

In Japan, *Termitomyces* spp. are found only in some parts of Okinawa Prefecture (Otani 1979; Otani and Shimizu 1981; Kinjo et al. 2005). Their host termite in Okinawa is *Odontotermes formosanus* (Yasuda et al. 2000; Sokabe et al. 2011). These fungi produce delicious mushrooms, but Japanese people are generally not familiar with them due to their restricted distribution and very short season.

In order to artificially produce *Termitomyces* mushrooms, we have begun an investigation of the nutritional requirements of these species. Revealing the cultural characteristics would be conducive to clarify symbiosis of the fungi and the termites. However, *Termitomyces* fungi distributed in Japan are generally shown to be very slow growing (Kinjo 2000). In this study, we collected many strains of *Termitomyces* in Japan and tried to select strains for nutritional analyses in combination with artificial media so as to determine optimal culture conditions.

5.2 Materials and Methods

Twenty-seven strains of *Termitomyces* were used for screening. All strains were isolated from fruiting bodies. Firstly, all the strains were identified as *Termitomyces eurrhizus* based on the morphological characteristics of their fruiting bodies (Otani 1979), and were confirmed to be the same species with DNA sequence analyses of internal transcribed spacer (ITS) regions 1 and 4(Kinjo 2012). They were collected in Okinawa Prefecture, Japan, in 2004, 2005, 2007, and 2009. Six strains (T1-6) were from the main island of Okinawa, nine strains (T7-15) were from Ishigaki Island, eleven strains (T-16-22 and T24-27) were from Iriomote Island, and one strain (T28) was from Kohama Island.

Eight media were used in this experiment:

1. Potato dextrose agar (PDA, Nissui Pharmaceutical Co. Ltd., Tokyo, Japan)
2. Half concentration of PDA (1/2 PDA)
3. Corn meal agar (CMA, Becton, Dickinson and Company, Sparks, USA)
4. Malt extract agar 1 (MA-1, malt extract (2%) [W/V], agar (2%) [W/V])
5. Malt extract agar 2 (MA-2, malt extract (2%) [W/V], peptone (0.5%) [W/V], agar (2%) [W/V])
6. Matsutake (EBIOS® [Asahi Food & Healthcare Co. Ltd., Tokyo, Japan] (0.5%) [W/V], glucose (2%) [W/V], agar (2%) [W/V])
7. Potato carrot agar (PCA)
8. Hay extract agar (HA)

All media were fashioned as φ 90-mm plates. The methods for PCA and HA preparations were as follows (Aoshima et al. 1983):

PCA: 20 g potato and 20 g carrot were grated and boiled in 1000 ml distilled water for 1 h. The filtrate was adjusted to 1000 ml with 2% (W/V) agar.

HA: 50 g hay (Timothy Glass [*Phleum pratense*]) in 1000 ml distilled water was autoclaved at 121 °C for 30 min. The filtrate was adjusted to 1000 ml with 0.2% (W/V) K_2HPO_4 and 2% (W/V) agar.

All the strains were pre-cultured on the same media, and 7-mm diameter pieces were taken with a cork borer and inoculated to the plate media. The plate media were cultured in the dark at 26 °C for a maximum 50 days. The diameter of the mycelium mats was measured at 5-day intervals in two vertical directions and was averaged. The growth rate was calculated as an average speed (mm/day) over the period from day 15–30 post-inoculation, when most of the strains showed a linear growth. When mycelium mats did not grow over 7-mm diameter on the pre-culture media by day 60 post-inoculation, these strains were not tested on those media. The tested strains were slow growing, and some of them exhibited frequent stagnation or sudden suppression of growth. In some cases, they died during cultivation. From these characteristics, the strains were rated as A, easy to handle; B, medium; and C, difficult to handle. The test was repeated three times for each medium ($n = 3$).

The mycelial growth rates were analyzed using an analysis of variance (ANOVA) and Tukey-Kramer tests for each medium, respectively ($p < 0.05$).

5.3 Results and Discussion

The growth rates of all the strains and handling rates are shown in Table 5.1. There was considerable variety in the growth of the strains on each medium. All the strains grew very slowly and were sensitive to contamination; some strains stopped growing suddenly and died. Most of the strains had dense and very short aerial hyphae on the mycelial mats, and some strains had few aerial hyphae. The surface of the mycelial mats was white in color in the early growing stage (Fig. 5.1a) but changed to pale brown after long-term cultivation. In addition, some strains often showed radial or irregular folds on their mycelial mats (Fig. 5.1b).

All the *Termitomyces* strains were kept on the PDA after collection, and we used PDA as the standard medium. Other than the PDA, only the matsutake medium showed sufficient growth in all strains. The other media showed poor growth in 1–3 strains. For the quantitative comparison, a 0.2 mm/day difference was set up as the criterion. Seven strains showed faster growth on 1/2 PDA than on PDA. However, there was no strain representing faster growth on the PDA than on 1/2 PDA. Furthermore, twelve strains displayed a higher growth rate on MA-1 than on MA-2. However, only one strain recorded faster growth on MA-2 than on MA-1. These results suggested that *T. eurrhizus* tended to prefer nutritionally poor media to rich media.

Eight strains exhibited faster growth on CMA among all the media, and the growth of 22 strains on CMA was faster than on PDA. The mycelial mats of many

Table 5.1 Mycelial growth rates of strains of *Termitomyces eurrhizus*

Growth rates (mm/day)*																			
Collecting place	Strains	PDA		1/2PDA		CMA		MA-1		MA-2		Matsutake		PCA		HA		Data range	Handling***
Main Island	T1	0.8	c	0.8	c	1.0	c	0.9	c	0.5	d	1.0	c	0.9	cd	1.3	c	0.5–1.3	B
	T2	1.1	b	1.0	bc	1.6	a	1.0	b	0.9	c	1.3	b	1.1	bc	1.6	b	0.9–1.6	C
	T3	1.1	b	1.1	bc	1.4	b	0.9	c	0.9	bc	1.4	b	1.1	bc	1.5	b	0.9–1.5	B
	T4	0.8	c	0.7	c	1.0	c	1.1	b	0.5	d	0.9	cd	1.1	bc	1.2	c	0.5–1.2	B
	T5	1.0	b	1.2	bc	1.2	bc	1.1	b	1.0	bc	1.1	c	0.6	d	0.5	d	0.5–1.2	B
	T6	1.0	b	1.0	bc	1.2	bc	0.7	c	0.9	c	1.0	cd	1.0	cd	1.7	ab	0.7–1.7	C
Ishigaki Island	T7	1.4	a	1.7	a	1.6	ab	1.4	a	0.7	d	1.6	a	1.7	a	1.1	c	0.7–1.7	C
	T8	1.0	b	1.1	bc	1.8	a	1.1	b	1.0	b	1.1	cd	1.1	bc	1.3	c	1.0–1.8	B
	T9	1.1	b	1.4	ab	1.3	b	1.1	ab	0.8	cd	1.4	b	1.1	bc	1.8	a	0.8–1.8	C
	T10	0.9	b	0.9	bc	1.2	bc	1.3	ab	1.2	b	1.3	b	1.1	bc	1.1	c	0.9–1.3	B
	T11	1.3	a	1.5	a	1.6	a	1.1	ab	0.7	cd	1.1	c	1.7	a	1.5	b	0.7–1.7	B
	T12	0.8	bc	0.8	c	1.5	ab	1.1	b	0.8	cd	0.9	d	1.5	a	0.8	d	0.8–1.5	C
	T13	1.1	b	1.2	bc	1.5	ab	1.2	ab	1.2	b	1.2	c	1.4	b	1.5	b	1.1–1.5	B
	T14	1.0	b	1.1	bc	–**		0.9	bc	0.9	bc	1.2	c	–**		1.4	b	0.9–1.4	B
	T15	0.7	c	–**		–**		–**		–**		0.7	d	–**		1.1	c	0.7–1.1	C

Iriomote Island	T16	1.3	ab	1.2	ab	1.7	a	0.6	c	0.6	cd	1.1	c	1.5	a	–**		0.6–1.7	c
	T17	0.9	bc	1.0	bc	0.9	c	0.9	c	0.9	c	1.1	c	1.0	cd	1.7	ab	0.9–1.7	A
	T18	1.2	b	1.1	bc	1.4	ab	1.3	ab	1.4	a	0.9	cd	1.1	bc	2.1	a	0.9–2.1	C
	T19	0.9	bc	0.9	bc	–**		1.1	b	0.7	cd	1.0	cd	–**		1.2	bc	0.7–1.2	C
	T20	0.8	c	0.9	bc	1.3	bc	0.9	bc	0.8	cd	1.1	c	1.2	bc	0.6	d	0.6–1.3	B
	121	1.4	a	1.6	a	1.5	ab	1.4	a	1.3	ab	1.4	ab	1.7	a	1.4	b	1.3–1.7	A
	T22	1.2	ab	1.3	ab	1.3	bc	1.1	b	0.9	bc	1.1	c	1.7	a	0.8	d	0.8–1.7	B
	T24	1.0	b	1.2	bc	1.2	bc	1.1	b	0.8	cd	1.1	cd	1.3	bc	1.4	b	0.8–1.4	C
	T25	1.5	a	1.6	a	1.8	a	1.4	a	1.4	a	1.7	ab	1.7	a	1.5	b	1.4–1.8	C
	T26	1.4	a	1.6	a	1.7	a	1.5	a	1.3	ab	1.7	a	1.9	a	2.3	a	1.3–2.3	A
	T27	1.1	b	1.1	bc	1.3	bc	1.1	b	1.0	bc	1.0	cd	1.0	bc	1.7	b	1.0–1.7	B
Kohama Island	T28	1.0	b	0.9	bc	1.3	bc	–**		–**		1.4	b	0.9	cd	1.3	b	0.9–1.4	C
	Data range	0.7–1.5		0.7–1.7		0.9–1.8		0.6–1.5		0.5–1.4		0.7–1.7		0.6–1.9		0.5–2.3			
	Mean	1.1		1.2		1.4		1.1		0.9		1.2		1.3		1.4			

The values labeled with different letters are significantly different by Tukey-Kramer ($p < 0.05$). $n = 3$.

*Averages of day 15–30 post-inoculation

**Did not grow over φ 7-mm by day 60 post-inoculation on the pre-culture media

***Handling rates: *A* easy to handle; *B* medium, *C* difficult to handle

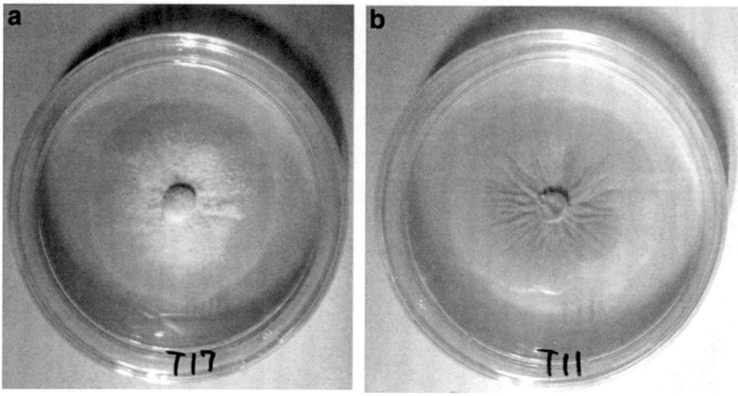

Fig. 5.1 (**a**) Strain T17 on PDA, 30 days post-inoculation; (**b**) strain T11 on PDA, 30 days post-inoculation

strains on CMA, however, were clearly thin in comparison with those on the other media. Seventeen strains showed the lowest growth rates on MA-2, and 15 strains demonstrated the fastest growth rates on HA among all the media. The mean growth rate on CMA and on HA was 1.4 mm/day; those were faster than on other media. But there were some strains that did not show sufficient growth on those media.

From these results, we selected four strains for further study. T3, from the strains of the main island of Okinawa, had low media specificity (0.9–1.5 mm/day growth rates on all media) and grew tractably. From the strains of Ishigaki Island, T11 was easily handled with similar growth rates to the other strains. In addition, this strain comparatively started to grow active after inoculation. T25 and T26 from the strains of Iriomote Island were also selected. These two strains grew very fast on any media. T25 was delicate and fragile but had low media specificity (1.4–1.8 mm/day growth rates on all media), whereas T26 was easy to handle. For the selection of media, the matsutake medium was evaluated as a good candidate for future experiments and keeping *Termitomyces* strains due to its availability to all the strains.

Acknowledgments We thank the late Dr. Kunio Tsunoda for his valuable suggestions on this study.

References

Aanen DK, Boomsma JJ (2007) In: Vega FE, Blackwell M (eds) Insect-fungal associations: ecology and evolution. Kyoritsu Shuppan, Tokyo, pp 237–260

Aoshima K, Tsubaki K, Miura K (eds) (1983) List of media composition. In: Methods of fungal research. Kyoritsu Shuppan, Tokyo, pp 398–399

Hyodo F, Tayasu I, Inoue T, Azuma J-I, Kudo T, Abe T (2003) Different role of symbiotic fungi in lignin degradation and food provision for fungus-growing termites (Macrotermitinae: Isoptera). Funct Ecol 17:186–193

Kinjo K (2000) Mushroom cultivated by termite. Shinrin Kagaku 30:20–25

Kinjo K (2012) Ohshiroaritake (*Termitomyces eurrhizus*). In: Yoshimura T, Itakura S, Iwata S, Ohmura W, Sugio K, Takematsu Y, Tokuda G, Matsuura K, Miura T (eds) Termites research in Japan: from biodiversity to control. Kaiseisha Press, Otsu, pp 91–94

Kinjo K, Promkemnerd A, Miyagi T (2005) *Termitomyces eurrhizus* collected from the main island of Okinawa. Jpn J Mycol 46:41–44

Nobre T, Rouland-Lefèvre C, Aanen DK (2011) Comparative biology of fungus cultivation in termites and ants. In: Bignell DE, Roisin Y, Lo N (eds) Biology of termites: a modern synthesis. Springer, Dordrecht, pp 193–210

Ohkuma M (2003) Termite symbiotic systems: efficient bio-recycling of lignocellulose. Appl Microbiol Biotechnol 61:1–9

Otani Y (1979) *Termitomyces albuminosus* (Berk.) Heim collected in Ishigaki Island of Ryukyu archipelago, Japan. Trans Mycol Soc Jpn 20:195–202

Otani Y, Shimizu D (1981) *Termitomyces clypeatus* Heim collected from Iriomote Island, Okinawa, Japan. Bull Natl Sci Mus Tokyo Ser B 7:131–134

Rouland-Lefèvre C, Lenoir F, Lepage M (1991) The role of the symbiotic fungus in the digestive metabolism of several species of fungus-growing termites. Comp Biochem Physiol 99A:657–663

Sokabe Y, Itakura S, Maegata M, Sakuratani Y (2011) A record of the butterflies and termites observed in Okinawa prefecture. Mem Fac Agric Kink Univ 44:131–137

Yasuda I, Nakasone Y, Kinjo K, Yaga S (2000) Morphology and distribution of termites in Ryukyu Islands and north and south Daito Islands. Jpn J Ent (N S) 3:139–156

Chapter 6
Volume Visualization of Hidden Gallery System of Drywood Termite Using Computed Tomography: A New Approach on Monitoring of Termite Infestation

S. Khoirul Himmi, Tsuyoshi Yoshimura, Yoshiyuki Yanase, Toshiyuki Torigoe, Masanori Akada, Motoko Ikeda, and Setsuo Imazu

Abstract In recent years, X-ray computed tomography (CT) has been developed as a reliable indirect method for insect studies. Here, we introduce the 3D volume visualization of hidden gallery system of drywood termites using computed tomography. Structural timbers infested by the western drywood termite, *Incisitermes minor*, with various dimensions were subjected into X-ray CT scanning analysis. The tomographic data were reconstructed by direct volume rendering using a volume graphic application. By rendering the tomographic data set, timber properties such as springwood, summerwood, and knots are able to be characterized; and by further specifying the opacity area, the nest-gallery systems inside the timbers are able to be visualized. This technique enables us to understand more biological features of this cryptic termite.

S.K. Himmi (✉)
Research Center for Biomaterials, Indonesian Institute of Sciences (LIPI), Jl. Raya Bogor km. 46 Cibinong, Bogor 16911, Indonesia
e-mail: khoirul_himmi@biomaterial.lipi.go.id

T. Yoshimura
Research Institute for Sustainable Humanosphere (RISH), Kyoto University, Gokasho, Uji-shi, Kyoto 611-0011, Japan

Y. Yanase
Graduate School of Agriculture, Kyoto University, Sakyo-ku, Kyoto 606-8502, Japan

T. Torigoe
Nara National Museum, 50 Noborioji-cho, Nara 630-8213, Japan

M. Akada
Kyushu National Museum, 4-7-2 Ishizaka, Dazaifu, Fukuoka 818-0118, Japan

M. Ikeda
Kyoto National Museum, 527 Chaya-cho, Higashiyama-ku, Kyoto 605-0931, Japan

S. Imazu
Nara University, 1500 Misasagi-cho, Nara-shi 631-0803, Japan

B. McLellan (ed.), *Sustainable Future for Human Security*,
https://doi.org/10.1007/978-981-10-5430-3_6

6.1 General Introduction

The structure of termite nests has sophisticated architecture and purpose-built functionality. Traniello and Leuthold (2000) reported that nests construction and development are related to feeding biology of termite's species and are important factors that influence termite's life biology and social evolution. Unique and particular interactions between termite individuals and surrounding environments create distinctive collective-responses and decision making (Deneubourg and Goss 1989). Therefore, the nest system is the expression of stigmergic behavior (Ladley and Bullock 2005) and coherent organization (Perna et al. 2008) of a termite colony.

The cryptic life of drywood termites makes it difficult to study their foraging behavior and to detect infestations inside wood. The western drywood termite, *I. minor*, is the most common structure-infesting drywood termite in the Southwestern United States (Cabrera and Scheffrahn 2001). Hagen (1858) characterized the species as the common drywood termite, *Kalotermes minor*, before Krishna (1961) reclassified it into genus *Incisitermes*. This termite is considered to be the most destructive drywood termite in the Western United States (USA) and one of the five most economically important termites in the United States (Su and Scheffrahn 1990). They adapt to the Mediterranean climate, a relatively dry climate with hot summers and little rainfall. Originally from the Southwestern United States and Northern Mexico (Light 1934), the distribution of this invasive species (Evans et al. 2013) has been reported to include Canada (Grace et al. 1991), China (Xie et al. 2001), Hawaii (Haverty et al. 2000), and to more than half of the prefectures in Japan (Harunari and Tomioka 2004; Indrayani et al. 2004; Yasuda et al. 2003; Yoshimura 2011).

Visual examination of drywood termite nests is technically very difficult. The dissection of wood, as presented on Fig. 6.1, is a common method to assess nest systems. However, this technique has major limitations, i.e., inability to examine the whole architecture and unique 3D connectivity of the nest gallery of drywood termites. The nest gallery of drywood termites consists of interconnected chambers and tunnel galleries (Fig. 6.1a). Some chambers may be filled with fecal pellets (Fig. 6.1b). Another major limitation of the dissection method is the inability to reveal important nest-gallery properties, such as total number of chambers, volume of chambers, and total length of nest gallery.

In the current study, we will introduce the 3D volume visualization of hidden gallery system of the western drywood termite, *Incisitermes minor*, using computed tomography. The term "volume visualization" is commonly used to describe all techniques that can be used to visualize volumetric data sets (Bosma 2000). An important field where these volume visualization methods can be applied is the visualization of volumetric data sets generated by three-dimensional scanners. We use a computerized X-ray imaging technique in which a narrow beam of X-rays is aimed at infested timbers to produce signals that are processed to generate two-dimensional (2D) tomographic images—or "image slices"—of those respective timbers. Once a number of successive image slices are rendered by the volume

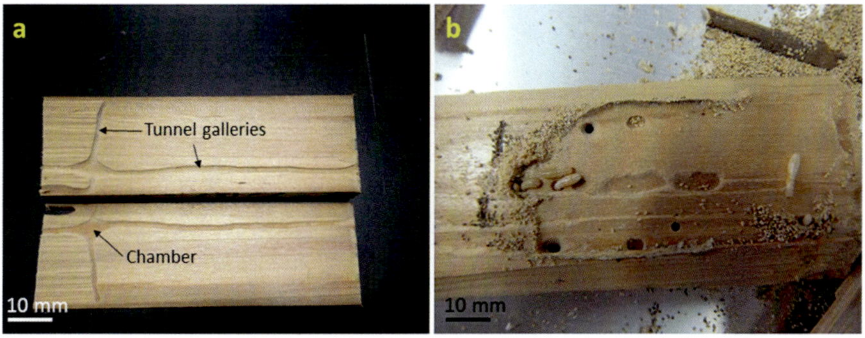

Fig. 6.1 Wood dissection as a visual assessment of drywood termite nest gallery cannot capture the whole architecture and 3D connectivity of nest gallery systems. (**a**) The nest gallery consists of interconnected chambers and tunnel galleries. (**b**) Some chambers, located beneath wood surface, are filled by fecal pellets

graphic application, they can be digitally "stacked" together to form a three-dimensional (3D) image. This technique is able to map the whole 3D connectivity of drywood termite nest gallery, which up till now is a technically difficult task to do. The technique is also able to estimate and monitor the extent of infestation inside the timbers, thus improve our level of understanding of this cryptic termite.

6.2 Materials and Methods

The structural timbers infested by the western drywood termite, *I. minor*, with various dimensions, were collected from a warehouse in Wakayama Prefecture, Japan, from 2009. The infested timbers were brought back to laboratory for further examination before being subjected into X-ray CT scanning analysis.

The scanning analysis was performed in Kyushu National Museum using YXLON X-Ray CT scanner (Y.CT Modular, YXLON International GmbH, Germany) and in Kyoto National Museum using Toshiba X-ray inspection systems (TOSCANER-32300 μ FD Series, Toshiba IT & Control Systems Corp., Japan). The tomographic data were reconstructed by direct volume rendering using a volume graphic application (VGStudio MAX 2.1, Volume Graphics GmbH, Germany).

Direct volume rendering uses an optical model to map data values to optical properties, such as color and opacity (Ikits et al. 2004). The classification method, as presented in Fig. 6.2, is the process of assigning opacity to a volume data. Classification is an essential part of the volume visualization (Praßni et al. 2010), as it enables the user to assign optical properties to individual parts of a volumetric data set. Volume data is stored as a stack of 2D image slices or as a single 3D texture object (Fig. 6.2). The term voxel in 3D image refers to an individual "volume element," similar to the terms pixel for "picture element" in 2D image.

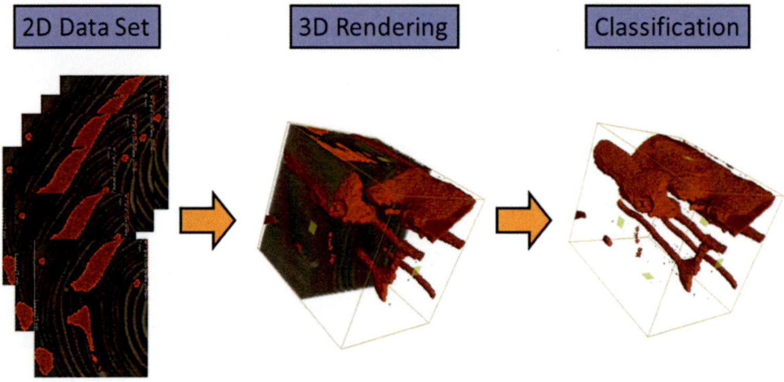

Fig. 6.2 The outline of 3D rendering of the tomographic data obtained from CT scan: The volumetric data is stored as a stack of 2D image slices or as a single 3D texture object, then the complete dataset was rendered by volume renderer tool on the software

Each voxel corresponds to a location in data space and has one or more data values associated with it. In this study, voxel data was shown in a histogram of pixel intensity value (opacity) which represented the absorptance intensity of the X-ray.

6.3 Results and Discussion

Figure 6.3 shows the result of direct volume rendering using the volume graphics application. The volume rendering technique delivered each voxel of the whole volume data with different opacity, as presented on the histogram of the image data (Fig. 6.3a). Different opacities were expressed with different frequencies of the image data value, resulting from the different absorptance intensities of the X-ray signal received by the timber. This technique enabled precise analysis of the timber properties by specifying the opacity of the voxel data on their gray value using the opacity transfer function (Praßni et al. 2010). The opacity transfer function is represented by the gray line plotted over the histogram.

In the histogram, an opacity curve specifies the transparency of the corresponding gray levels; the lower the line, the more transparent the corresponding gray value is rendered. The shape of the opacity curve can be adjusted by adding control points. By rendering the complete data set (Fig. 6.3a), timber properties such as springwood, summerwood, and knots can be characterized, and by further specifying the opacity area (Fig. 6.3b), the gallery systems inside the timber can be visualized.

Figure 6.4 presents the isosurface volume rendering technique. The isosurface is defined as the surface in a continuous area of the volume data at which the value equals to a predefined isovalue (Bosma 2000). Direct volume rendering (in Fig. 6.3) was performed by generating images of a 3D volumetric data set without explicitly extracting geometric surfaces from the data. On the other hand, isosurface rendering was conducted by dragging the isosurface line on the histogram of the data set

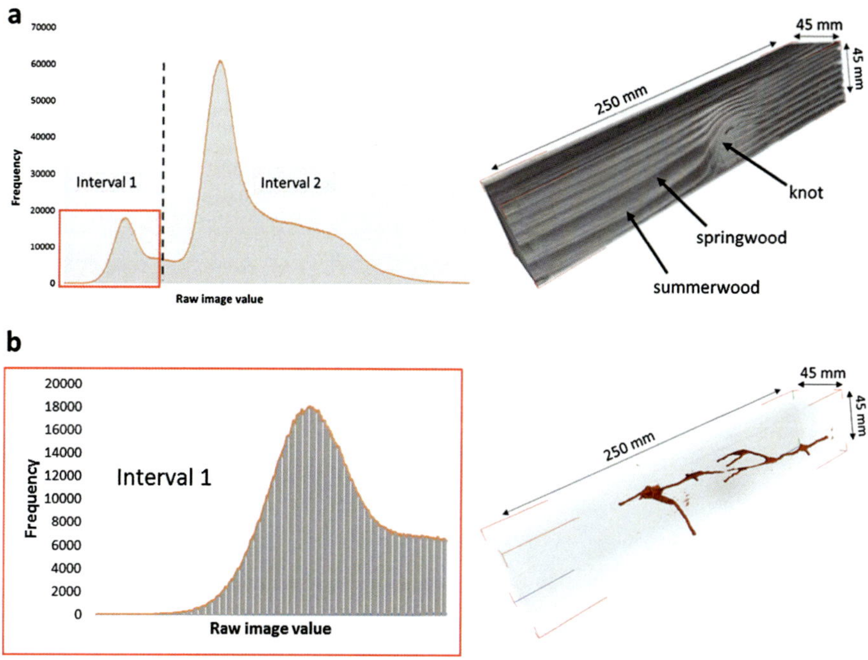

Fig. 6.3 The 3D rendering technique using a volume graphic application. (**a**) The histogram of CT image data and 3D visualization in volume renderer mode. (**b**) Visualization of the nest gallery inside the timber created by extracting interval 1 of the volumetric data set and by creating an independent volumetric data set

either to the left or to the right, in order to decrease or increase, respectively, the gray level at which the "isovalue" was rendered. Various densities of physical properties of the timber generated different data values. By specifying a certain value, the isosurface will render all of the voxels with values that are bigger than the specified isovalue. Therefore, we were able to specify the density profile of the timber and to show the respective parts, such as the summerwood parts (Fig. 6.5), and to omit the other parts of the timber with lower isovalues.

By applying isosurface volume rendering, density profile of the timber can be mapped; and by combining it with 3D volume visualization of nest-gallery system of drywood termite, important behavior on how drywood termite forages and extends the nest gallery inside the timber can be observed. Himmi et al. (2016) suggested that *I. minor* showed self organization and continual adaptation toward the nest environment. Self organization is the accummulative dynamic and complex collective behaviors and interactions among termite individuals to maintain the nest-gallery systems. Self organization is expressed on the disposal of fecal pellets to maintain networks and connectivities; the fortification behavior by sealing exposed holes using cement pellets. Continual adaptation of the nest environment was demonstrated by the anatomical selectivity in establishing the nest on less

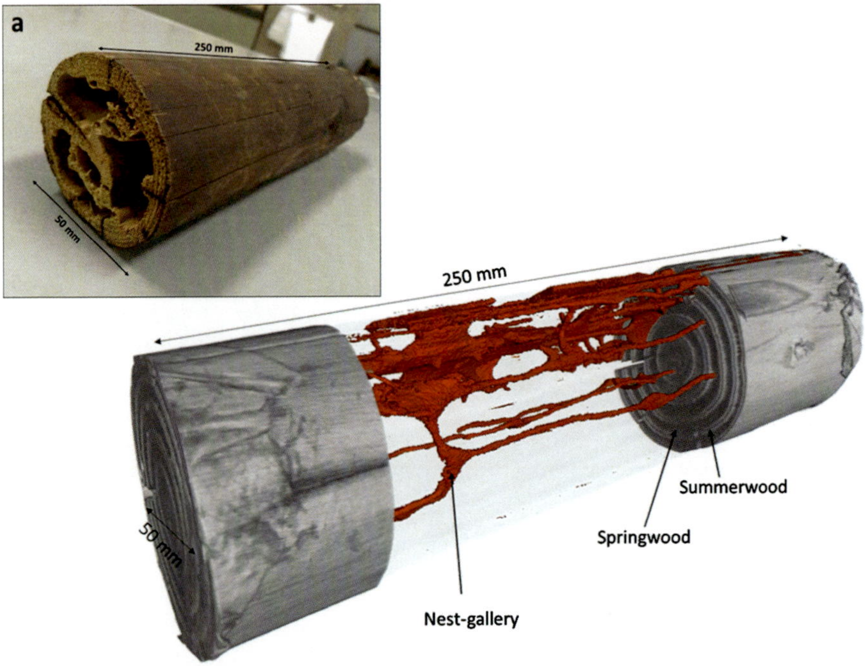

Fig. 6.4 The 3D isosurface volume rendering: By specifying an isovalue on the image data, the isosurface will render all the voxels with bigger values. Therefore, only the parts of the timber that are correlated with those voxels are visualized

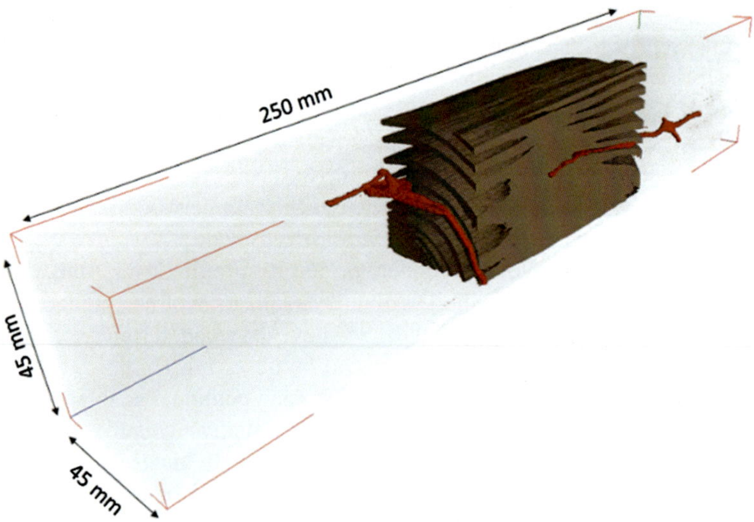

Fig. 6.5 The 3D isosurface volume rendering: The summerwood parts of annual growth rings can be visualized, while the springwood parts, which have lower density value, can be omitted

dense and more favorable wood areas inside timbers. Other species of drywood termites, *Cryptotermes domesticus* (Evans et al. 2005) and *C. secundus* (Evans et al. 2007), have also been reported to have ability to differentiate between food sources using vibrational signals. Therefore, drywood termites are able to identify the wood and to modulate their foraging behavior (Evans et al. 2005, 2007; Indrayani et al. 2007; Inta et al. 2007) based on foraging efficiency to maximize and to efficiently use food resources as measured by energy and time spent.

6.4 Conclusion

At the practical level, infestation of drywood termites is very difficult to locate, control, and manage, as the colony lives entirely inside the wood; and is often hidden from view. The critical challenges of the inspection process are how to identify the existence and the extent of the infestation. X-ray CT can lead us to a better understanding of the nesting biology, nest-gallery system, and foraging behavior of this economically important pest.

References

Bosma MK (2000) Iso-surface volume rendering: speed and accuracy for medical applications. University of Twente, Enschede

Cabrera BJ, Scheffrahn RH (2001) Western drywood termite, *Incisitermes minor* (Hagen) (Insecta: Isoptera: Kalotermitidae). Univ Fla IFAS Ext:1–7

Deneubourg JL, Goss S (1989) Collective patterns and decision-making. Ethol Ecol Evol 1:295–311. doi:10.1080/08927014.1989.9525500

Evans TA, Lai JC, Toledano E, McDowall L, Rakotonarivo S, Lenz M (2005) Termites assess wood size by using vibration signals. Proc Natl Acad Sci U S A 102:3732–3737. doi:10.1073/pnas.0408649102

Evans TA, Inta R, Lai JCS, Lenz M (2007) Foraging vibration signals attract foragers and identify food size in the drywood termite, *Cryptotermes secundus*. Insect Soc 54:374–382. doi:10.1007/s00040-007-0958-1

Evans TA, Forschler BT, Grace JK (2013) Biology of invasive termites: a worldwide review. Annu Rev Entomol 58:455–474. doi:10.1146/annurev-ento-120811-153554

Grace JK, Cutten GM, Scheffrahn RH, Kevan DKM (1991) First infestation by *Incisitermes minor* of a Canadian building (Isoptera: Kalotermitidae) Sociobiol 18:299-304

Hagen HA (1858) Monographie der Termiten, Part 2. Linnaea Entomologica 12:1–342

Harunari M, Tomioka Y (2004) Occurrence and control of exotic termite, *Incisitermes minor* (Hagen), in Yokohama (in Japanese). House Househ Insect Pests 26:107–113

Haverty MI, Woodrow RJ, Nelson LJ, Grace JK (2000) Cuticular hydrocarbons of termite of the Hawaiian islands. J Chem Ecol 26:1167–1191

Himmi SK, Yoshimura T, Yanase Y, Mori T, Torigoe T, Imazu S (2016) Wood anatomical selectivity of drywood termite in the nest-gallery establishment revealed by X-ray tomography. Wood Sci Technol 50:631–643. doi:10.1007/s00226-016-0800-x

Ikits M, Kniss J, Lefohn A, Hansen C (2004) Volume rendering techniques. Addison-Wesley Professional, Canada

Indrayani Y, Yoshimura T, Fujii Y, Yanase Y, Okahisa Y, Imamura Y (2004) Survey on the infestation of houses by *Incisitermes minor* (Hagen) in Kansai and Hokuriku areas. Jpn J Appl Entomol Zool 15:261–268

Indrayani Y, Yoshimura T, Yanase Y, Fujii Y, Matsuoka H, Imamura Y (2007) Observation of feeding behavior of three termite (Isoptera) species: *Incisitermes minor,* Coptotermes formosanus, and *Reticulitermes speratus*. Sociobiology 49:121–134

Inta R, Lai JC, Fu EW, Evans TA (2007) Termites live in a material world: exploration of their ability to differentiate between food sources. J R Soc Interface 4:735–744. doi:10.1098/rsif.2007.0223

Krishna K (1961) A generic revision and phylogenetic study of the family Kalotermitidae (Isoptera). Bull Am Mus Nat Hist 122:303–408

Ladley D, Bullock S (2005) The role of logistic constraints in termite construction of chambers and tunnels. J Theor Biol 234:551–564. doi:10.1016/j.jtbi.2004.12.012

Light SF (1934) The distribution and biology of the common drywood termite, *Kalotermes minor*: distribution and means of identification. In: Kofoid CA (ed) Termites and termite control. University of California Press, Berkeley, pp 210–216

Perna A, Jost C, Couturier E, Valverde S, Douady S, Theraulaz G (2008) The structure of gallery networks in the nests of termite *Cubitermes spp.* revealed by X-ray tomography. Naturwissenschaften 95:877–884. doi:10.1007/s00114-008-0388-6

Praßni J-S, Ropinski T, Mensmann J, Hinrichs K (2010) Shape-based transfer functions for volume visualization. In: Pacific Visualization Symposium (PacificVis), IEEE, 2010. IEEE, Piscataway, pp 9–16. doi:10.1109/PACIFICVIS.2010.5429624

Su NY, Scheffrahn RH (1990) Economically important termites in the United States and their control. Sociobiol 17:77–94

Traniello JFA, Leuthold RH (2000) Behavior and ecology of foraging in termites. In: Abe T, Bignell DE, Higashi M (eds) Termites: evolution, sociality, symbioses, Ecology. Kluwer Academic Publishers, Dordrecht, pp 141–168

Xie Y, Li Z, Gregg WP, Li D (2001) Invasive species in China: an overview. Biodeversity and Conservation 10:1317–1341

Yasuda I, Kinjo K, Yaga S (2003) A new record of *Incisitermes minor* (Hagen) from Okinawa island (in Japanese). Jpn J Entomol 6:103–104

Yoshimura T (2011) Strategies towards the integrated management of the invasive dry-wood termite, *Incisitermes minor* (in Japanese). Mokuzai Gakkaishi 57:329–339

Part III
Biomass and Microbiology in Tropical Regimes

Chapter 7
The Potency of Fenton-Polyaluminum Chloride for Black Liquor Treatment

Ajeng Arum Sari, Anis Kristiani, Hendris Hendarsyah Kurniawan, and R. Irni Fitria Anggraini

Abstract Black liquor is one of the main by-products of the pretreatment process in bioethanol production from empty oil palm fruit bunches, with a high chemical oxygen demand (COD) and low dissolved oxygen (DO). The effect of $FeSO_4$ as a coagulant and $FeSO_4$–H_2O_2 for Fenton on the degradation of black liquor was examined. This study also identified the ability of a fungus to decolorize black liquor wastewater after the Fenton process and on original black liquor. One percent ferrous sulfate decolorized 84% of black liquor with a concentration of 30,000 ppm under the coagulation method. By adding H_2O_2 and $FeSO_4$ through the Fenton process, decolorization of the original black liquor was approximately 52%. Combining Fenton and polyaluminum chloride decolorized black liquor up to 90% in 33 min, whereas *Coriolus versicolor* decolorized 54% and 75% Fenton-treated black liquor and original black liquor after 15 days, respectively.

7.1 Introduction

Indonesia is the largest palm oil producer in the world. One ton crude palm oil generates 1.1 ton oil palm empty fruit bunches (OPEFB) as lignocellulosic biomass waste. OPEFB content consists of cellulose (29.9%), hemicellulose (18.6%), lignin (27.6%), and others (Sari et al. 2014). Cellulose and hemicellulose from OPEFB can be converted to ethanol. Research Center for Chemistry, Indonesian Institute of Sciences, has been developing technology for bioethanol production that consists of four steps: pretreatment, hydrolysis of cellulose to produce sugars, fermentation of sugars to ethanol, and distillation to obtain purified ethanol.

The pretreatment process has several purposes such as removing lignin and hemicellulose, increasing surface area, and fractionating amorphous cellulose (Kristiani et al. 2016). Several options for pretreatment process that are commonly used are physical/mechanical, chemical, biological, and a combination of these

A.A. Sari (✉) • A. Kristiani • H.H. Kurniawan • R.I.F. Anggraini
Research Center for Chemistry, Indonesian Institute of Sciences, Kawasan Puspiptek Serpong, Tangerang Selatan 15314, Indonesia
e-mail: ajen001@lipi.go.id; ajeng_as@yahoo.co.id

B. McLellan (ed.), *Sustainable Future for Human Security*,
https://doi.org/10.1007/978-981-10-5430-3_7

types (Mosier et al. 2005; Taherzadeh and Karimi 2008; Alvira et al. 2010). The chemical pretreatment using sodium hydroxide (NaOH) aims to alter the structure of cellulosic biomass by removing lignin. After that, the cellulose becomes more accessible to the enzymes that have a role for converting it into fermentable sugars (Alvira et al. 2010). During alkali pretreatment, lignin is solubilized and/or decomposed in the aqueous phase resulting in a soluble fraction containing lignin degradation products (Sudiyani et al. 2016). It is called black liquor.

In the Research Center for Chemistry, Indonesian Institute of Sciences, 600 kg oil palm empty fruit bunches are used to produce 76.46 kg bioethanol. It generated 3000 l black liquor. It has black color with high chemical oxygen demand (COD) and low dissolved oxygen (DO) and also may be potentially harmful to aquatic ecosystems if it discharges directly into waters without any particular treatment system (Sari et al. 2015). Black liquor consists of chlorinated compounds, suspended solids, phenolics, and resin along with lignin.

Treatment for black liquor from bioethanol production specifically is still limited. However, this black liquor has similar characteristics with black liquor from pulp and paper process. Several technologies using physical, chemical, and biological methods can be applied to treat black liquor (Table 7.1).

The technology needed for greater removal of COD through lignin structure destruction is high oxidation from OH radical systems, such as *advanced oxidation processes* (AOPs), photocatalytic systems, and Fenton reactions (Cundy et al. 2008). Fenton technology is commonly used for wastewater treatment because it can destroy organic pollutants, remove BOD/COD, and remove odor and color (Araujo et al. 2002). The major advantage of Fenton reactions, using $FeSO_4.7H_2O$, is high reaction yields, easy to obtain, and short times, and it is economical, thereby making it a more cost-effective process, unlike other treatments. Reactions related to the hydroxyl radical and regeneration ion Fe^{2+} are as follows:

$$Fe^{2+} + H_2O_2 \rightarrow Fe^{3+} + {}^{\bullet}OH + {}^{-}OH \quad k = 76.5\ \mathrm{L\ mol^{-1}\ s^{-1}}$$

$$Fe^{3+} + H^2O^2 \rightarrow FeOOH^{2+} + H^+ \quad K_{eq} = 3 \cdot 10^{-3}$$

$$FeOOH^{2+} \rightarrow HO_2^{\bullet} + Fe^{2+} \quad k = 2.70 \cdot 10^{-3}\,\mathrm{s^{-1}}$$

$$Fe^{3+} + HO_2^{\bullet} \rightarrow Fe^{2+} + O_2 + H^+ \quad k < 2.00 \cdot 10^{-3}\ \mathrm{L\ mol^{-1}\ s^{-1}}$$

However, conventional methods, such as coagulation, have also shown satisfactory results when removing black liquor (Irfan et al. 2013). Many alternative coagulants can be used too, such as polyaluminum chloride (PAC), lime, alum, ferric chloride, and ferrous sulfate. Unfortunately, several chemicals used in these methods generate sludge after the process.

Moreover, fungal treatments using white-rot fungus can be one option for combination treatments to degrade black liquor wastewater, because they secrete ligninolytic enzymes which attack pollutants and reduce the toxicity of said pollutants (Sari and Tachibana 2012). *Phanerochaete chrysosporium*, *Trametes*

Table 7.1 Technologies for black liquor and other pulp and paper wastewater

No.	Technology	Removal efficiency	Reference
1	Coagulation (aluminum chloride)-flocculation (starch-g-PAM-g-PDMC)	Turbidity, 95.7%; lignin, 83.4%	Wang et al. (2011)
2	Electrocoagulation	COD, 95%	Shankar et al. (2013)
3	Fenton-like ($FeCl_3$-H_2O_2)	Color, 55.57%; lignin, 37.49%	Sari et al. (2017)
4	Adsorption (ion exchange resin and granular activated carbon)	DOC, 72–76%	Ciputra et al. (2010)
5	Ultrafiltration Ultrafiltration plus dissolved air flotation	TOC, 54%; TSS, 100%; color, 88% TOC, 65%; TSS, 100%; color, 90%	De Pinho et al. (2000)
6	Reverse osmosis	BOD, 88%; COD, 89%	Dube et al. (2000)
7	Solar photo-Fenton (Fe^{2+}/H_2O_2/UV)	COD, 90%	Lucas et al. (2012)
	Solar photocatalytic degradation ($nTiO_2$/UV)	COD, 75%; TSS, 80%	Ghaly et al. (2011)
8	Aerated lagoon in pH 7.3	Ammonia, 67%	Bryant et al. (1997)
9	Sequencing batch reactor (SBR)	COD, 90%	Reid and Simon (2000)
10	Fungi + solar photo-Fenton	COD, 90%; DOC, 90%	Fernandes et al. (2014)
11	Continuous anaerobic treatment	COD, 48–80%	Poggi-Varaldo et al. (1996)
12	Anaerobic biological granular activated carbon	COD, 50%; color, 50%	Jackson-Moss et al. (1992)
13	Fungal specie (*Penicillium* sp.)	AOX, 50%	Taseli and Gokcay (1999)
14	Fungal specie (*Pleurotus ostreatus*)	Lignin, 77%; BOD, 76.8%; COD, 60%; color, 80%	Choudhury et al. (1998)

versicolor, and *T. elegans* are recognized for their ability to degrade lignin and black liquor wastewater (Mehna et al. 1995).

The general aim of the present investigation was to determine the degradation of black liquor wastewater by using a combined chemical and biological treatment. The objective was to examine the effect of $FeSO_4$ as a coagulant and $FeSO_4$–H_2O_2 for Fenton reactions on the degradation of black liquor. This study also identified the ability of the fungus to decolorize black liquor wastewater.

7.2 Material and Methods

7.2.1 *Materials*

Black liquor wastewater was obtained during the pretreatment process in bioethanol production in the Research Center for Chemistry, Indonesian Institute of Sciences. It consists of an alkaline aqueous solution that contains organic and inorganic solids and has a distinctive dark coloration. The characteristics of black liquor are shown in Table 7.2. The research scheme is explained in Fig. 7.1.

7.2.2 *Degradation of Black Liquor by Coagulation*

The coagulant used in this work for decolorization of black liquor was ferrous sulfate, with a concentration of 0.1–2%. 150 ml of the effluent sample with a concentration of 30,000 ppm (v/v) was taken in a glass beaker, mixed vigorously for 1 min by using a jar test at 200 rpm, and after that slowly mixed for 2 min at 50 rpm. The effluent sample was left steady for 30 min for sedimentation.

Table 7.2 Characteristics of black liquor wastewater

Parameters	Amount
pH	12.4
Chemical oxygen demand (COD)	149,000 mg/L
Total suspended solid (TSS)	17,200 mg/L
Color	Black

Source: measurement

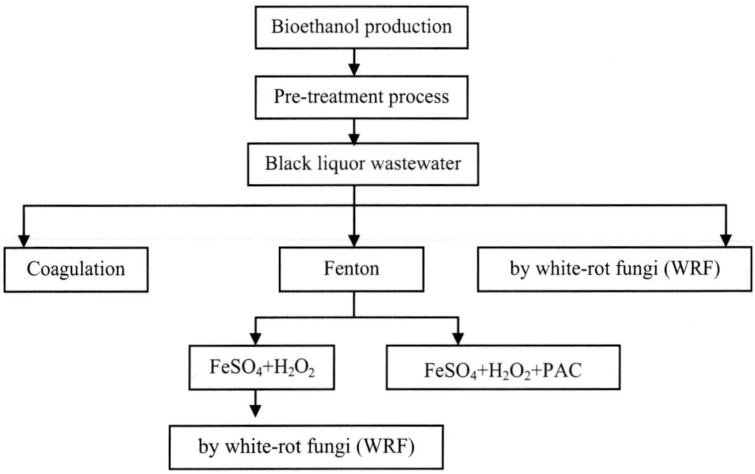

Fig. 7.1 Research scheme for the degradation of black liquor

7.2.3 Degradation of Black Liquor by Fenton

A hydroxyl radical was generated by the Fenton reagent (0.25 g $FeSO_4$ + 50 ml H_2O_2). After that, 150 ml of black liquor was added to this reagent. This solution was mixed strongly for 2 min by using a jar test at 200 rpm, thereafter mixed slowly for 5 min at 50 rpm, and left steady for 24 h for sedimentation. For the next experiment, 2.5 mg polyaluminum chloride (PAC) was added into the Fenton solution. The procedure was the same as the previous explanation.

7.2.4 Degradation of Black Liquor by White-Rot Fungus

The white-rot fungus used in this study, *Coriolus versicolor*, is preserved in the Research Center for Chemistry, Indonesian Institute of Sciences. The sample used for this study was maintained on malt extract agar (MEA) medium at 25 °C for 7 days. For decolorization test, an agar plug (Ø 5 mm) from 7-day-old mycelium of strain was inoculated in plates (Ø 90 mm) containing Czapek-Dox with 1% Remazol Brilliant Blue R (RBBR) (Sari and Tachibana 2012). The inoculated plates were incubated at 25 °C for 7 days and the level of decolorization was assessed.

An agar plug of fungi on a maintained medium agar was tested for growth on AG medium containing 30,000 ppm of black liquor (Da Re and Papinutti 2011). In the first layer, the AG medium made consisted of agar, 15 g L^{-1}; glucose, 10 g L^{-1}; KH_2PO_4, 0.5 g L^{-1}; K_2HPO_4, 0.6 g L^{-1}; $CuSO_4$, 0.4 mg L^{-1}; $MnCl_2$, 0.09 mg L^{-1}; Na_2MoO_4, 0.02 mg L^{-1}; $FeCl_3$, 1 mg L^{-1}; $ZnCl_2$, 3.5 mg L^{-1}; and thiamine hydrochloride, 0.1 mg L^{-1} at pH 4.5. 15 g L^{-1} of agar mixed with black liquor 30,000 ppm was added on the second layer. The inoculated plate was incubated at 25 °C for 7 days, and the level of growth was assessed by measuring the radial extension of mycelium. The growth or decolorization percentage was calculated by measuring the growth disc diameter of each fungus, according to the formula

$$\%\ \text{growth} = \left(\frac{\phi_c - \phi_s}{\phi_c}\right) \times 100\%$$

where

ϕ_s = fungal diameter in treated sample (with black liquor)
ϕ_c = fungal diameter in control (without black liquor)

The AG liquid medium for this experiment was made with the same nutrient components as the agar medium. Black liquor was used to replace distilled water. There are two types of experiment. First, the decolorization of the original black liquor by fungus was determined. Second, the ability of fungus to decolorize black liquor that has already been degraded by Fenton was determined. Three agar plugs of the fungus were inoculated in 20 ml of autoclaved AG liquid medium in 100 ml

Erlenmeyer flasks and then incubated for 3, 7, 15, 21, and 30 days. After incubation, samples were filtrated to obtain the supernatant that was measured in the next experiment.

7.2.5 Analytical Methods

Decolorization was determined by the change in the absorbance of black liquor, determined at 365 nm with a UV-Vis spectrophotometer (Optizen 2120 UV). Black liquor without treatment was used as a control. The percentage of decolorization was calculated as follows:

$$\text{Decolorization}\ (\%) = \left(1 - \frac{C}{C_0}\right) \times 100$$

where C_0 is the initial black liquor concentration and C is the final black liquor concentration.

COD and TSS analyses were performed by standard methods from APHA/AWWA/WEF (2005). The mycelial of the fungus was separated from the medium using filter paper and then dried at 55 °C for 28 h. The weight of mycelial on the filter was measured to determine the mycelial mass of fungus. The pH before and after degradation was measured by using a pH meter.

Physicochemical characterization of black liquor after treated by Fenton and Fenton-PAC was analyzed. Scanning electron microscope (SEM) images were obtained using the Scanning electron microscope (SEM) ZEISS, and the SEM images were taken by applying 10 kV voltage with various magnification times for the observation of the surface. To identify various functional groups present in the black liquor after treated by Fenton and Fenton-PAC, FTIR analysis was carried out using a FTIR spectrophotometer (Shimadzu, Prestige-21). Samples were pressed into small discs and placed in the IR cell. All infrared spectra were collected data with a resolution of 4 cm^{-1} at the range of 4000–400 cm^{-1}.

7.3 Results and Discussion

7.3.1 Degradation by Coagulation

Coagulant ferrous sulfate doesn't have the ability to decolorize the original black liquor, though the black liquor was diluted up to 30,000 ppm. The highest decolorization of black liquor, 84%, was obtained under 1% ferrous sulfate (Fig. 7.2).

The pH of the wastewater was reduced to 6.8 after the addition of the coagulant. The presence of an iron-based coagulant can separate organic compounds from the

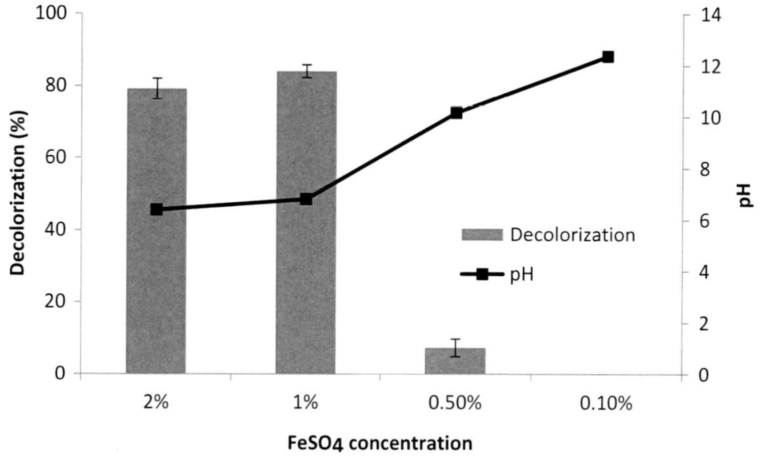

Fig. 7.2 Decolorization of 30,000 ppm black liquor by using coagulation method

Fig. 7.3 Decolorization of black liquor by using Fenton

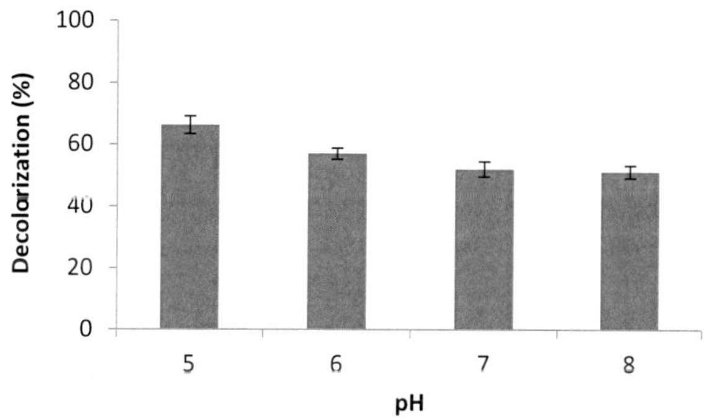

wastewater by the destabilization of colloids and suspended matter, allowing particle collision and growth of floc (Sahu and Chaudhari 2013).

7.3.2 *Degradation of Black Liquor by Fenton*

Original black liquor was used in this experiment. By adding H_2O_2 and less $FeSO_4$, decolorization of the original black liquor was approximately 52%. It is known that addition of H_2O_2 influences the decomposition of complex organic compounds due to an increase in the available OH radicals by the Fenton reaction (El-Dein et al. 2001). After adding H_2O_2, the pH was decreased, which attributed to the fragmentation of organic material into organic acids (Taherzadeh and Karimi 2008) (Fig. 7.3).

Table 7.3 Degradation of black liquor by using Fenton-PAC

Parameter	Fenton + PAC	
	Time	Degradation (%)
Decolorization	33 min	90
TSS		70
COD		55

The H_2O_2 in the Fenton reagent attacks the phenolic groups of lignin in black liquor because these functional groups have a very high affinity for reactive oxygen species, thus decolorizing the black liquor wastewater (Araujo et al. 2002). Since decolorization at pH 7 was still approximately 60%, this wastewater was further treated by using adding polyaluminum chloride (PAC) or a biological method with white-rot fungus.

By adding PAC, degradation was significantly improved, because PAC may slow the rate of hydroxide precipitation upon dilution, can be maintained for a longer time, and can be more strongly adsorbed at surfaces (Sanghi et al. 2006). Carboxylic and phenolic groups from lignin reacted with metal cations of PAC, thus increasing the degradation of black liquor (Table 7.3).

Studies have shown that during Fenton's oxidation, color removal of the dye was faster than COD removal (Kang and Chang 1997). The reason could be due to the formation of stable intermediate products, which require a longer time for further oxidation.

7.3.3 Degradation by White-Rot Fungus

Coriolus versicolor was able to decolorize RBBR on agar medium contaminated with RBBR and grow in malt extract agar medium contaminated with black liquor. The decolorization of RBBR was noticeable on day 2 and achieved a maximum of 100% decolorization on 9 cm diameter of petri dish by 10 days as seen in Fig. 7.4. A colorless halo around the microbial growth indicated ligninolytic enzymes were produced by WRF (Kiiskinen et al. 2004). A change in color from dark blue to orange indicated disappearance of RBBR to breakdown of the chromophoric compound in the dye (Hadibarata et al. 2012). Because of that reason, *C. versicolor* was assumed to have the ability to degrade several xenobiotic compounds including black liquor. Furthermore, this strain was tested in agar medium contaminated with black liquor. On agar medium contaminated with 30,000 ppm of black liquor, *C. versicolor* was able to grow by approximately 90% on the seventh day.

In a liquid medium, this strain has different abilities to decolorize original black liquor in the two types of experiment (Fig. 7.5).

However, decolorization of Fenton-treated black liquor was lower than original black liquor. After a degradation of 52% by using Fenton, black liquor was further decolorized by approximately 54% in 15 days. On the other hand, original black

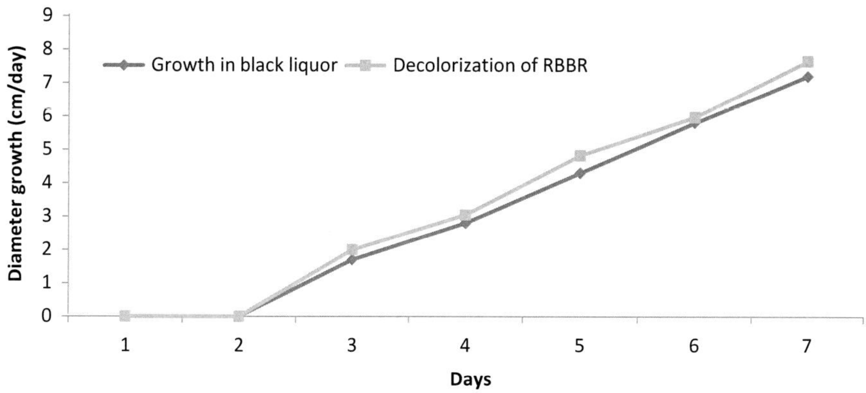

Fig. 7.4 Ability of *C. versicolor* to decolorize RBBR and grow in black liquor agar media

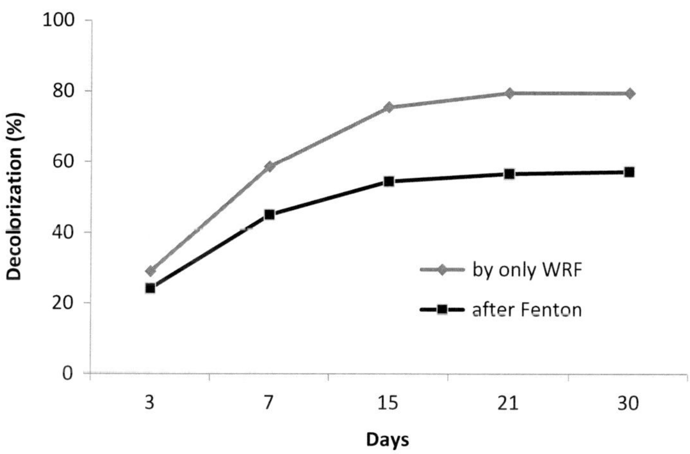

Fig. 7.5 Decolorization of black liquor by *Coriolus versicolor*

liquor was decolorized up to 75%. This was potentially caused by a high concentration of $FeSO_4$ and H_2O_2 in the medium, thereby causing inhibition of enzyme secretion of the fungus. Surprisingly, on day 15, the mycelia dry weight of fungus in the black liquor medium was higher than in the nutrient medium without black liquor (data not shown). It means that the growth of the fungus was not affected by the presence of black liquor and Fenton compounds. Decolorization by fungus was similar to the decolorization by Fenton + PAC method, though it took longer. *Pleurotus sajor-caju* has been used to treat wastewater from secondary treatment of a bleached kraft *Eucalyptus globulus*. The reduction of relative absorbance at 465 nm reached 72% (Freitas et al. 2009). The advantages of the biological treatment are an absence of sludge and a decrease in toxicity because the enzyme produced by fungi can break down the structure of black liquor. However, the

Fig. 7.6 SEM images of (**a**) black liquor-Fenton and (**b**) black liquor-Fenton-PAC

limitations of fungal treatment are that it is unable to be used under an extreme environmental condition such as high pH and oxygen defects, the high glucose requirement, and the necessity of longer incubation time because of slow metabolism of the microorganisms (Kamali and Khodaparast 2015).

Through the SEM measurements, the microtopography of the samples could be clearly observed as shown in SEM images of Fig. 7.6. In Fig. 7.6a, the surface of black liquor-Fenton is rough and irregular. After being combined with PAC, the surface becomes more distinct, and the folding structure can be clearly observed in Fig. 7.6b. They are formed by the coating of PAC on the surface. The particles found have a rigid structure. The color of the center of the nanoparticle is darker, which is ascribed to the existence of Fenton. In contrast, due to the coating of PAC, the color of the edge of the particle becomes lighter. The particle size increased twice approximately from 0.8 to 1.6 μm after treatment.

FTIR spectra indicate that during treatment processes, there were no significant changes of functional groups occurred in the pattern of wavenumber spectra as shown in Fig. 7.7. The bands of IR absorbance of black liquor at the wavenumber range of 1800–3500 cm^{-1} indicate the presence of OH group bound via hydrogen bond. The existence of C-H group aromatic at wavenumber of 700–800 cm^{-1} was also observed. The wavenumber of 1600–2900 cm^{-1} indicates the presence of a carboxylic acid group (OH), and the absorbance in this area also strengthened the indication of aromatic (C=C) dominance (Pari et al. 2003). The pattern of IR spectra indicates that the intensity of absorbance at 600–800 cm^{-1} decreased as the effect of the Fenton treatment.

Physicochemical methods have shown the ability to remove suspended solids and recalcitrant pollutants from black liquor. The coagulation, Fenton, and fungal treatments were evaluated regarding their capacity for decolorization of black liquor. The combination of several technologies such as AOPs and coagulation can be considered as an effective way to decolorize black liquor and produce smaller quantity of sludge. Combining Fenton and polyaluminum chloride gave the highest decolorization of black liquor, up to 90% in 33 min. This method is preferable to be applied to a great scale. However, many factors should be

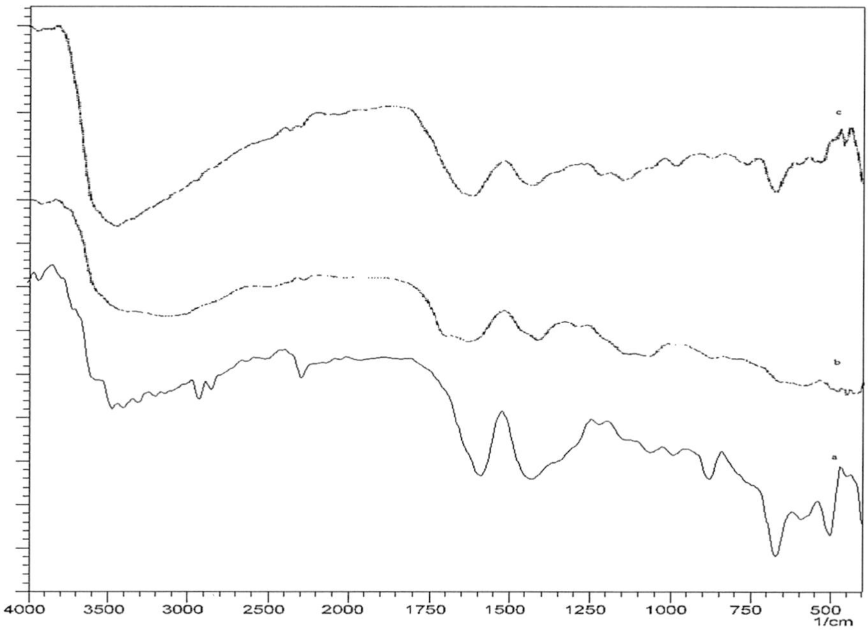

Fig. 7.7 FTIR spectra of (*a*/*bottom*) black liquor (*b*/*middle*) treated by PAC and (*c*/*top*) treated by Fenton

considered in continuous processes such as loading rate, pump pressure, and maintenance. Further, a life cycle analysis of this technology in a pilot scale should also be considered. After selecting the system boundary, SimaPro 7.3.3 can be used to select impact categories, characterization models, and optional (normalization, grouping, and weighting) elements of the life cycle impact assessment (LCIA) according to ISO 14040.

7.4 Conclusion

It was concluded that original black liquor wastewater can be effectively decolorized by using two methods: Fenton + PAC and Fenton + *Coriolus versicolor*. By adding H_2O_2 and $FeSO_4$ through the Fenton process, decolorization of the original black liquor was approximately 52%. Combining Fenton and polyaluminum chloride decolorized black liquor up to 90% in 33 min, whereas *Coriolus versicolor* decolorized 54% and 75% Fenton-treated black liquor and original black liquor after 15 days, respectively. These results open the possibility for the use of combined physiochemical and biological treatment for black liquor wastewater from bioethanol process. However, the utilization of Fenton sludge still needs further consideration.

Acknowledgments This research was partly supported by the Program Insentif Riset Sistem Inovasi Nasional (Insinas) – Ministry of Research, Technology and Higher Education Fiscal Year 2015–2017. We thank Mr. Nick V for language editing and proofreading this manuscript.

References

Alvira P, Tomás-Pejó E, Ballesteros M, Negro MJ (2010) Pretreatment technologies for an efficient bioethanol production process based on enzymatic hydrolysis: a review. Bioresour Technol 101:4851–4861

APHA-AWWA-WEF (2005) Standard methods for the examination of water and wastewater, 21st edn. American Publishing Health Association, American Water Works Association and Water Environment Federation, New York

Araujo E, Rodrigues-Malaver AJ, Gonzalez AM, Rojas OJ, Penaloza N, Bullon J, Lara MA, Dmitrieva N (2002) Fenton's reagent-mediated degradation of residual kraft black liquor. Appl Biochem Biotechnol 97:91–103

Bryant CW, Barkley WA, Garett RM, Gardner FD (1997) Biological nitrification of kraft wastewater. Water Sci Technol 35(2–3):147–153

Choudhury S, Sahoo N, Manthan M, Rohela RS (1998) Fungal treatment of pulp and paper mill effluents for pollution control. J Ind Pollut Control 14(1):1–13

Ciputra S, Antony A, Phillips R, Richardson D, Leslie G (2010) Comparison of treatment options for removal of recalcitrant dissolved organic matter from paper mill effluent. Chemosphere 81:86–91

Cundy AB, Hopkinson L, Whitby RLD (2008) Use of iron-based technologies in contaminated land and groundwater remediation: a review. Sci Total Environ 400:42–51

Da Re V, Papinutti L (2011) Black liquor decolorization by selected white-rot fungi. Appl Biochem Biotechnol 165(2):406–415

De Pinho MN, Minhalma M, Rosa MJ, Taborda F (2000) Integration of flotation/ultrafiltration for treatment of bleached pulp effluent. Pulp Pap Can 104(4):50–54

Dube M, McLean R, MacLatchy D, Savage P (2000) Reverse osmosis treatment: effects on effluent quality. Pulp Pap Can 101(8):42–45

El-Dein M, Libra A, Wiesmann JA (2001) Kinetics of decolorization and mineralization of the azo dye reactive black 5 by hydrogen peroxide and UV light. Water Sci Technol 44(5):295–301

Fernandes L, Lucas MS, Maldonado MI, Oller I, Sampaio A (2014) Treatment of pulp mill wastewater by *Cryptococcus podzolicus* and solar photo-Fenton: a case study. Chem Eng J 245:158–165

Freitas AC, Ferreira F, Costa AM, Pereira R, Antunes SC, Goncalves F, Rochasantos TAP, Diniz MS, Castro L, Peres I, Duarte AC (2009) Biological treatment of the effluent from a bleached kraft pulp mill using basidiomycete and zygomycete fungi. Sci Total Environ 407:3282–3289

Ghaly MY, Jamil TS, El-seesy IE, Souaya ER, Nasr RA (2011) Treatment of highly polluted paper mill wastewater by solar photocatalytic oxidation with synthesized nanoTiO_2. Chem Eng J 168:446–454

Hadibarata T, Yusoff ARM, Kristanti RA (2012) Decolorization and metabolism of anthraquinone-type dye by laccase of white rot fungi *Polyporus sp.* S133. Water Air Soil Pollut 223:933–941

Irfan M, Butt T, Imtiaz N, Abbas N, Khan RA, Shafique A (2013) The removal of COD, TSS and colour of black liquor by coagulation–flocculation process at optimized pH, settling and dosing rate. Arab J Chem. http://dx.doi.org/10.1016/j.arabjc.2013.08.007

Jackson-Moss CA, Maree JP, Wotton SC (1992) Treatment of bleach plant effluent with the biological granular activated carbon process. Water Sci Technol 26(1–2):427–434

Kamali M, Khodaparast Z (2015) Review on recent developments on pulp and paper mill wastewater treatment. Ecotoxicol Environ Saf 114:326–342

Kang SF, Chang HM (1997) Coagulation of textile secondary effluents with Fenton's reagent. Water Sci Technol 36(12):215–222

Kiiskinen LL, Ratto M, Kruus K (2004) Screening for novel laccase-producing microbes. J Appl Microbiol 97(3):640–646

Kristiani A, Effendi N, Styarini D, Aulia F, Sudiyani Y (2016) The effect of pretreatment by using electron beam irradiation on oil palm empty fruit bunch. Atom Indonesia 42(1):9–12

Lucas MS, Peres JA, Amor C, Prieto-rodriguez L, Maldonado MI (2012) Tertiary treatment of pulp mill wastewater by solar photo-Fenton. J Hazard Mater 225–226:173–181

Mehna A, Bajpai P, Bajpai PK (1995) Studies on decolorization of eluent from a small pulp mill utilizing agriresidues with *Trametes versicolor*. Enzym Microb Technol 17:18–22

Mosier N, Wyman C, Dale B, Elander R, Lee YY, Holtzapple M, Ladisch M (2005) Features of promising technologies for pretreatment of lignocellulosic biomass. Bioresour Technol 96:673–686

Pari G, Sofyan K, Syafii W (2003) Buchari. The properties of charcoal from the black liquor of the soda pulping of rice straw. Jurnal Teknologi Hasil Hutan 16(2):70–79

Poggi-Varaldo HM, Estrada-Vazquez C, Fernandez-Villagomez G, Esparza-Garcia F (1996) Pretreatment of black liquors spills effluent. In: Proceedings of the industrial waste conference, West Lafayette, vol 51, pp 651–661

Reid TK, Simon A (2000) Feasibility study of sequencing batch reactor technology treating high strength foul condensate for methanol reduction. In: Tappi international environmental conference and exhibit, vol 1, pp 185–191

Sahu OP, Chaudhari PK (2013) Review on chemical treatment of industrial waste water. J Appl Sci Environ Manag 17(2):241–257

Sanghi R, Bhattacharya B, Dixit A, Singh V (2006) Ipomoea dasysperma seed gum: an effective natural coagulant for the decolorization of textile dye solutions. J Environ Manag 81:36–41

Sari AA, Tachibana S (2012) Muryanto. Correlation of ligninolytic enzymes from the newly-found species *Trametes versicolor* U97 with RBBR decolorization and DDT degradation. Water Air Soil Pollut 223(9):5781–5792

Sari AA, Kristiani A, Tachibana S, Sudiyani Y, Abimanyu H (2014) Mechanisms and optimization of oil palm empty fruit bunch as a pre-grown source for white-rot fungus to degrade DDT. J Environ Chem Eng 2:1410–1415

Sari AA, Kurniawan HH, Nurdin M, Abimanyu H (2015) Decolorization of black liquor wastewater generated from bioethanol process by using oil palm empty fruit bunches. Energy Procedia 68:254–262

Sari AA, Muryanto M, Waluyo J, Wiloso EI (2017) Life cycle assessment comparison of Fenton and Fenton-like for wastewater treatment in second generation bioethanol production. In: Proceedings of the 2nd ILCAN conference series on Life cycle assessment, vol 2, pp 27–31

Shankar R, Singh L, Mondal P, Chand S (2013) Removal of lignin from wastewater through electro-coagulation. World J Environ Eng 1:16–20

Sudiyani Y, Triwahyuni E, Muryanto M, Burhani D, Waluyo J, Sulaswaty A, Abimanyu H (2016) Alkaline pretreatment of sweet sorghum bagasse for bioethanol production. J Renew Energy Dev 5(2):113–118

Taherzadeh MJ, Karimi K (2008) Pretreatment of lignocellulosic wastes to improve ethanol and biogas production: a review. Int J Mol Sci 5:337–353

Taseli B, Gokcay CF (1999) Biological treatment of paper pulping effluents by using a fungal reactor. Water Sci Technol 40(11–12):93–99

Wang J, Chen Y, Wang Y, Yuan S, Yu H (2011) Optimization of the coagulation-flocculation process for pulp mill wastewater treatment using a combination of uniform design and response surface methodology. Water Res 45:5633–5640

Chapter 8
Pretreatment of Sweet Sorghum Bagasse Using EFB-Based Black Liquor for Ethanol Production

Muryanto and Ajeng Arum Sari

Abstract Lignocellulose-based bioethanol provides one potential alternative energy. Pretreatment is one of the steps in the bioconversion of lignocellulose material. Pretreatment also contributes the largest cost in the bioethanol production and produces black liquor as a wastewater which provides environmental impacts. In order to improve the cost-effectiveness of bioethanol production, the black liquor from empty fruit bunch (EFB) pretreatment was collected and used for pretreatment of sweet sorghum bagasse (SSB). The pretreatment process was conducted in a 5-liter reactor at 150 °C. The result by using black liquor was compared with pretreatment by using NaOH solution. The delignification of SSB by using black liquor in 30 and 60 min pretreatment time was 61.57% and 55.87%, respectively. The ethanol production after the SSF process reached 45.06 gl^{-1} and 46.49 gl^{-1}.

8.1 Introduction

Lignocellulose bioethanol is an alternative fuel or oxygenated additive to the current fossil fuels which has become of interest in recent decades. Biofuel based on lignocellulose is considered as a critical issue to solve the conflict between food and energy resources for fuel production (Cassman and Liska 2007). The advantages of lignocellulose bioethanol are that the raw material is abundant and does not require very fertile land, it can reduce greenhouse gas emissions, and it can reduce the food price due to decreasing competition with food crops (Harshvardhan and Pal 2015). Lignocellulose refers to plants or biomass with a primary cellulose and lignin component; its source can be from bioenergy crops and also from lignocellulosic waste. The characteristics of the ideal energy crop is high-yield production,

Muryanto (✉) • A.A. Sari
Research Center for Chemistry, LIPI, Kawasan Puspiptek Serpong, Tangerang, Selatan 15314, Indonesia
e-mail: moer_yanto@yahoo.com

B. McLellan (ed.), *Sustainable Future for Human Security*,
https://doi.org/10.1007/978-981-10-5430-3_8

low energy input, and low cost to produce (McKendry 2002). Sweet sorghum is a multipurpose crop that can be cultivated under a wide range of environmental conditions for simultaneous production of grain, sugar juice, and lignocellulose bagasse (Wu et al. 2011). Sorghum has high biomass yield which can be harvested at up to 58.3–80.5 tons of fresh stem per hectare (Chambergo et al. 2015). It may be a promising crop for bioethanol production. Cai et al. (2013) reported that adding sorghum feedstock in existing ethanol production can fulfill the volume of renewable and cellulosic bioethanol in term of analysis life cycle energy and greenhouse gas emission. Sweet sorghum stems contain soluble carbohydrates like glucose and xylose that are the raw materials of first-generation bioethanol and also have insoluble carbohydrate (cellulose and hemicelluloses) in bagasse that can be the source of second-generation bioethanol (Saibi et al. 2013).

In terms of chemical composition, the sweet sorghum bagasse (SSB) predominantly contains cellulose, hemicelluloses, and lignin. Cellulose, the main important material for bioethanol, is a polysaccharide composed of D-glucose subunits, linked by β-1,4 glycosidic bonds (Hendriks and Zeeman 2009). Cellulose has a tendency to form microfibrils through inter- and intramolecular hydrogen bonding. Hemicellulose is a mixture of polysaccharides, composed almost entirely of sugars such as glucose, mannose, xylose, and arabinose. Xylose is monomer of sugar that is predominately hemicellulose. Lignin is further linked to both hemicellulose and cellulose forming a physical seal around the latter two components that is an impenetrable barrier preventing penetration of solution and enzymes.

The conversion of lignocelluloses to ethanol includes four steps: the first is pretreatment, consist of mechanical pretreatment, chemical pretreatment, biological pretreatment, and combinations of these. Chemical pretreatment can use acid or alkaline for the solvent. The second step is the hydrolysis of cellulose into glucose that can be used for chemical or enzymatic hydrolysis. Enzymatic hydrolysis was favorable due to it high-yield production (Chambergo et al. 2015). The cellulase enzyme is the mixture of the following enzymes, exoglucanase, endoglucanase, cellobiohydrolase, and β-glucosidase (Rizk and Antranikian 2012). The next step is fermentation to convert monomeric sugar (glucose, xylose) to ethanol. Commonly, fermentation was conducted using species of yeast, *Saccharomyces cerevisiae*. This yeast only utilizes glucose to convert to ethanol (Ohgren et al. 2006; Hahn-Hägerdal et al. 1994). Thus, there remains some xylose product that can be used as a coproduct in bioethanol processing. Genetic modification has been studied to metabolic engineer *Saccharomyces cerevisiae* that can convert both glucose and xylose into ethanol but has not yet been fully successful. Purification of bioethanol is needed to produce bioethanol of fuel grade. The common technologies that are used to purify bioethanol are distillation and dehydration. Bioethanol purification systems must be operated continuously for high performance with minimum energy requirement (Ismael et al. 2008). The last step is residue treatment from the process. Generally, the residue comes from the pretreatment process and fermentation process. This waste can be recycled, incinerated, or converted to other uses such as animal feed or fertilizer (Chambergo et al. 2015).

Alkaline pretreatment is one of the chemical pretreatments that can be used for pretreatment of lignocellulose materials (Fan et al. 1987). The goal of any pretreatment is delignification and altering or removing structural and compositional obstacles to hydrolysis, to improve the rate of enzyme digestibility, and to increase the yield of fermentable sugars from substrates (Mosier et al. 2005). Delignification must be done to destroy the lignin structure and to give an optimum result in bioconversion of lignocelluloses to fermentable sugar (Taherzadeh and Karimi 2008).

Although the yields have improved, higher environmental impact in second-generation ethanol scenarios are mainly related to high NaOH consumption for delignification prior to hydrolysis, being the most impacting parameter in global warming (Dias et al. 2012). Pretreatment processes using a large amount of NaOH also increase the operational costs of ethanol production. Generally, the black liquor produced from alkaline pretreatment should be discharged and not be reusable. Moreover, black liquor contains valuable chemical compounds, such as lignin, which could be recovered. Black liquor also contains high NaOH concentration. As a consequence, the black liquor is a serious pollutant that must be treated, and high treatment costs can make the production process uncompetitive. The high content of NaOH in black liquor has the potential to be used as a solvent in the pretreatment process. According to this, technological improvements, such as NaOH recycling, are necessary in this process for the environmental sustainability of second-generation ethanol production.

Previous studies have examined recycling black liquor for empty fruit bunch (EFB)'s pretreatment (Muryanto et al. 2015). The results showed that black liquor can be used as a pretreatment solvent, but NaOH addition was required to maintain pretreatment effectiveness. Xu et al. (2012) used switchgrass-derived black liquor for corn stover pretreatment process without NaOH addition to reduce the overall chemical cost. The current study aims to use the black liquor from NaOH pretreatment of EFB at 150 °C (2.5 M NaOH, 30 min, 4 bars) for SSB's pretreatment. Pretreatment of SSB using EFB's black liquor is expected to reduce costs of production of bioethanol and also minimize the wastewater from the process.

8.2 Material and Methods

8.2.1 Material

Sweet sorghum bagasse (SSB) used in this experiment was obtained from a plantation in Pasuruan East Java, Indonesia, after the extraction of sugar juice. Physical pretreatment was conducted as the first step to reduce the size to below 3 mm and then the material was stored in a plastic bag. The black liquor was collected from pretreatment of EFB using 10% NaOH solution, 150 °C, 4 bar and

30 min. The commercial enzymes Cellic® Ctec2 and Cellic® Htec2 from Novozyme and the commercial dry yeast (*Saccharomyces cerevisiae*) were applied for the simultaneous saccharification process. The vendor reported that the activity of CTec2 was 144 FPU (Filter Paper Unit)/ml enzyme, whereas HTec2 was 240 CBU (cellobiose unit)/ml.

8.2.2 Chemical Pretreatment

Chemical pretreatment was performed by heating 250 g of SSB (10% moisture content) with 1250 ml black liquor in a stirred reactor. The pressure was controlled at 4 bar; the reactor temperature was set at 150 °C with reaction time 30 and 60 min. After pretreatment, the solid fraction was washed and neutralized then dried at 50–60 °C overnight. Characterization of SSB was measured before and after pretreatment. Cellulose, hemicellulose, and lignin contents were analyzed based on a method from the National Renewable Energy Laboratory (NREL) (Ruiz and Ehrman 1996). SSB (300 mg, dry weight) was subjected to acid hydrolysis for lignin, cellulose, and hemicellulose content analysis. After hydrolysis, acid-insoluble lignin (AIL) was weighed using Sartorius BS224S and acid-soluble lignin (ASL) was measured using Spectrophotometry UV/Vis Spectrophotometer Optizen 2120 UV at 205 nm. Total lignin was obtained from the sum of AIL and ASL. On the other hand, after hydrolysis, cellulose and hemicellulose were measured by HPLC Waters e2695.

8.2.3 Simultaneous Saccharification and Fermentation (SSF)

The SSF process was conducted in 250 ml Erlenmeyer with 15% (w/v) or 15 g dry weight basis as a substrate, and then 75 ml of 0.05 M citrate buffer pH 4.8 was added. Citrate buffer and substrate were sterilized using an autoclave. The enzyme and buffer citrate were added to the sample until the total volume was 100 ml. Two kinds of enzymes, CTec2 and HTec2, were added with the ratio 5:1. The CTec2 loading was 30 FPU g^{-1} pretreated EFB. The yeast loading amount was 1% b/v (1 g). The SSF was conducted in a shaker incubator for 72 hr at a temperature of 32 °C and 150 rpm agitation.

8.2.4 Analysis

Samples (1 ml) were withdrawn from the SSF medium every 24 h. Glucose, xylose, and ethanol were determined using high-performance liquid chromatography (HPLC) Waters e2695. The HPLC system was equipped with an Aminex HPX-87H column (Bio-Rad, Richmond, CA, USA) and operated at 65 °C with 5 mM H_2SO_4 as the mobile phase at a flow rate of 0.6 ml min^{-1} along with 25 min acquisition time.

8.3 Results and Discussion

The chemical compositions of untreated sweet sorghum bagasse and empty fruit bunch are reported in Table 8.1. Before pretreatment, total carbohydrate in the SSB was 48.81%, and total lignin fraction was 21.39% of the dry biomass. The carbohydrate content in this SSB was lower than other species that are used by Li et al. (2010), but the lignin content in SSB was 32% lower than of EFB's in Sudiyani et al. research (2013). Even the same type of biomass, the chemical composition would be different; many factors can affect it such as environment conditions (pH, temperature, sunlight, and fertilizer), growth period, harvesting time, and storage conditions (Bradshaw et al. 2007; Weimer and Springer 2007). SSB physical characteristics of being softer than EFB allows it to be pretreated by lower NaOH concentration.

The cellulose content of SSB after pretreatment was increased with both NaOH solution and black liquor. The composition of pretreated SSB with different pretreatment solution is listed in Table 8.2. The cellulose contents of SSB after pretreatment using black liquor and NaOH solution were 62.01% and 66.08% (30 min and 60 min) and 52.74% and 67.37% (4% NaOH and 6% NaOH), respectively. Reused black liquor as the pretreatment solution increased the cellulose content with the result being between the pretreatment using 4% NaOH and 6% NaOH solution. The performance of black liquor in the pretreatment process of SSB could be attributed to two factors. The first factor is the high pH of black liquor that reached 12.97. This indicated that the NaOH concentration in black liquor is quite high. Our previous research mentions that the NaOH content in the black liquor is around 76% from the first process (Muryanto et al. 2015). NaOH concentration is the important key in the alkaline pretreatment. The second factor is the presence of carbohydrates in the initial black liquor collected from EFB pretreatment. This result is in line with the Xu result's that using black liquor for pretreatment of corn stover (Xu et al. 2012). Generally, if the black liquor is not recycled for biomass pretreatment, its carbohydrates will be burnt for heat generation or discarded due to the difficulty in their separation and recovery. This process could reduce chemical utilization in the bioethanol process. Reusing the black liquor for pretreatment solvent could also minimize the amount of black liquor discharged into the

Table 8.1 Chemical composition of untreated EFB

Composition	EFB (%) Sudiyani et al. (2013)	SSB (%)
Cellulose	37.26	34.09
Hemicellulose	14.62	14.72
Lignin	31.68	21.39
Ash	6.69	4.91

Table 8.2 Chemical composition of untreated and pretreated bagasse

No	Pretreatment condition	Cellulose (%)	Hemicellulose (%)	Lignin (%)	Ash(%)
1	BL, 30 min, 150 °C	62.01	25.8	8.22	0.18
2	BL, 60 min, 150 °C	66.08	27.10	9.44	0.64
3	NaOH 4%, 30 min, 130 °C	52.74	18.95	17.82	1.56
4	NaOH 6%, 30 min, 130 °C	67.37	13.5	5.09	0.28

environment so that the bioethanol process becomes more environmentally friendly. The black liquor can be reused several times until lignin content was saturated and must be the treatment. The lignin in black liquor could separate by coagulation process and could be process into activated carbon (Amriani et al. 2015; Fu et al. 2013).

Hemicellulose content also decreased after pretreatment because pretreatment with NaOH can also dissolve hemicellulose as has been done by Wan et al. (2011) on soybean straw. Besides influenced by the alkali solution, the pretreatment process was also influenced by time (Sjostrom 1993). In this study, we also carried out pretreatment process using the variation of time (30 min and 60 min) in the black liquor process. These pretreatment conditions showed that increasing the time of processing can increase the cellulose and hemicelluloses content.

One of the purposes of alkaline pretreatment is to reduce lignin content called delignification. Delignification can improve the ability of cellulase enzymes in the saccharification process. Millet et al. (1976) and Kang et al. (2013) reported that the performance of the enzyme increased by 14% to 55% on hardwood when the lignin content was reduced from 24–55% to 20%. The lignin reduction of pretreated SSB using black liquor at 30 min and 60 min reached 61.57% and 55 87%, respectively (Fig. 8.1). This result was compared with pretreatment using NaOH solution. In the previous study, black liquor was quite effective to remove lignin in the EFB with NaOH addition resulting in 59.30% delignification (Muryanto et al. 2015). So comparatively, the reuse of black liquor without the addition of NaOH can lower lignin content of SSB up to 50–60%. The high pH levels in black liquor indicated NaOH content was still high so that it can reduce lignin in SSB. But the delignification result was still slightly lower compared with 6% NaOH solution. According to Xu et al. (2012), the ability of black liquor to reduce lignin was caused by the presence of organic compounds in the black liquor such as phenol. Phenol has been proven effective in degrading lignin from wheat straw (Zacchi et al. 1988).

Pretreatment of SSB with NaOH solution increases cellulose and reduces the lignin content; the result is in line with Goshadrou result's (Goshadrou et al. 2011).

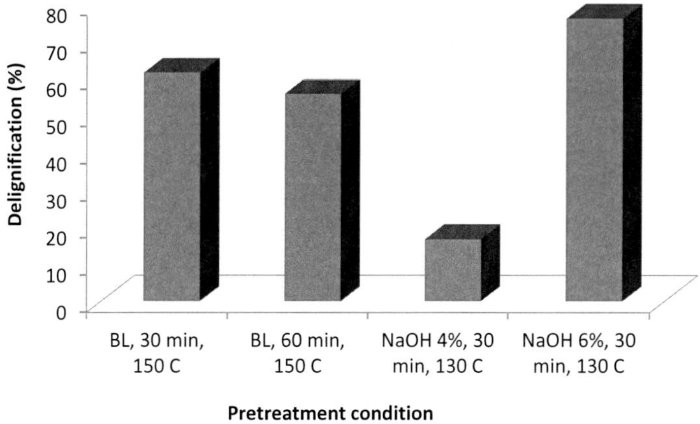

Fig. 8.1 Delignification of SSB

According to Goshadrou (Goshadrou et al. 2011), the spectra from FTIR analysis of pretreated bagasse were changed, the absorbance at about 3350 cm^{-1} (related to cellulose) was increased, and absorbance at 1218 cm^{-1} (related to lignin) was decreased.

The main objective of the pretreatment is to degrade the physical structure of the biomass and eliminate lignin and hemicellulose to increase percentage and accessibility of cellulose fraction. It is also used to change the structure biomass to make the enzyme more accessible (Ioelovich 2015). The alkaline pretreatment contributes to enhancing the enzymatic digestibility of the pretreated substrate. The main factors influencing the enzyme digestibility are the content of cellulose and lignin in the pretreated substrate. Ioelovich (2015) reported that the yield of sugar after enzymatic hydrolysis depends on the difference between the content of cellulose and lignin in the substrate. Increasing the difference of cellulose minus lignin value can increase the yield of sugar in the hydrolysis process.

The pretreated SSB was converted to bioethanol by enzymatic hydrolysis and fermentation. Enzymatic hydrolysis converted cellulose into monomer sugar (glucose) by cellulose enzymes. As side products, glucose and cellobiose also could be inhibitors in the hydrolysis process. The fermentation process was conducted by *Saccharomyces cerevisiae* to change glucose after fermentation into ethanol. *S. cerevisiae* converted glucose to ethanol by *Embden-Mayerhof pathway*, also known as the glycolysis process. In glycolysis, one mole of glucose is converted to two moles of pyruvate. Pyruvate in *S. cerevisiae* could be converted to ethanol by *Pyruvate decarboxylase* (PDC) and *alcohol dehydrogenase* (ADH) enzymes (Pronk et al. 1996). Hydrolysis and fermentation of lignocellulosic material can be conducted using several process technologies. The process could be carried out sequentially in separate reactors as separate hydrolysis and fermentation (SHF). The other alternative is simultaneous saccharification and fermentation (SSF) in which the enzyme and yeast simultaneously inserted at the beginning of the process

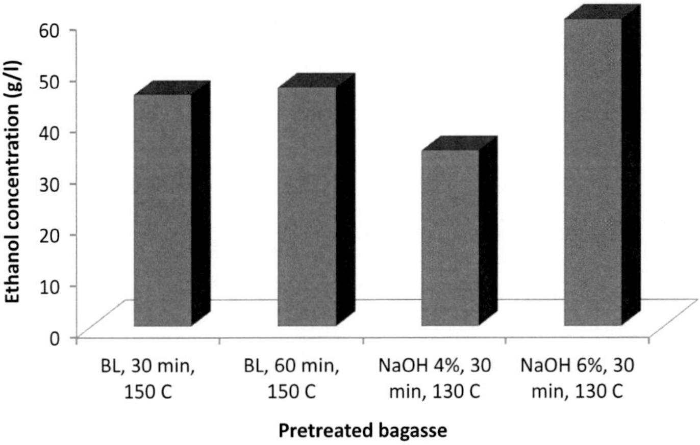

Fig. 8.2 Ethanol production from pretreated SSB

and the reactions are conducted in one reactor. SSF is an efficient process because it reduces the number of reactors and can obtain high and rapid ethanol production (Wingren et al. 2003, Cantarella et al. 2004, Dahnum et al. 2015).

Figure 8.2 shows the ethanol concentration in the variation of pretreated SSB. The pretreated SSB was converted to bioethanol using SSF process at pH 4.8 and temperature process 32 °C. The pH was kept using buffer citrate to prevent glycerol production. According to Li et al. (2010), the production of glycerol was higher at pH 6.0 than pH 4.8. It was shown in earlier work that the biosynthesis of glycerol could be stimulated in higher pH (Ingledew 1999). The ethanol concentration from pretreated SSB by using black liquor at 30 min and 60 min were 45.06 gl^{-1} and 46.49 gl^{-1}, respectively. This result is in line with ethanol concentration produced by pretreated SSB in the work conducted by Li et al. (2010) using ammonia fiber extraction (AFEX) that reached 42.3 gl^{-1}.

The ethanol concentration from pretreated SSB using black liquor was higher than the ethanol concentration resulting from pretreated SSB using 4% NaOH solution which only reached 34.19 gl^{-1} but still lower than pretreated SSB using 6% NaOH solution. The summary data from the SSF process of pretreated SSB in variations of the pretreatment process was shown in Table 8.3. The ethanol yield concentration based on cellulose content in the SSB was 0.48, 0.47, 0.43, and 0.59 gg^{-1} cellulose at pretreatment using black liquor (30 min and 60 min) and using NaOH solution (4% and 6%), respectively. The ethanol yield based on pretreated SSB using black liquor over either 30 min or 60 min was close to ethanol concentration from pretreated SSB using 6% NaOH solution.

As expected, pretreatment was necessary to convert the cellulose efficiently in the bagasse to fermentable sugars using cellulose enzymes. Additionally, the pretreatment could increase the cellulose fractions. The results of this work showed that the black liquor pretreatment could produce the ethanol yield and productivity as well as sodium hydroxide solution. The reuse of black liquor as a pretreatment solvent could reduce the NaOH consumption and could also minimize wastewater

Table 8.3 The yield of ethanol in simultaneous saccharification and fermentation (SSF) of pretreated sweet sorghum bagasse

Variable	Max. ethanolconcentration (gl^{-1})	Yield ethanolbased on cellulose ($g\ g^{-1}$ cellulose)[a]	Yield ethanolbased on pretreated SSB ($g\ g^{-1}$pret-SSB)[b]
BL, 30 min, 150 C	45.07	0.48	0.30
BL, 60 min, 150 C	46,50	0.47	0.31
NaOH 4%, 30 min, 130 C	34.19	0.43	0.23
NaOH 6%, 30 min, 130 C	59.77	0.59	0.40

[a]yield ethanol based on cellulose = max ethanol/cellulose fraction in SSB
[b]yield ethanol based on SSB = max ethanol/dry weight SSB (150 gl^{-1})

discharge to the environment so that the process would be more environmentally friendly. The lower chemical requirements in the process increase efficiency and also reduce cost of production. Pretreatment is one of the major production costs in the lignocellulosic bioethanol due to high energy and chemical consumption. When the chemical consumption was reduced, this means that costs can also be reduced for the pretreatment process and the lignocellulosic bioethanol can be more competitive with the other energy sources.

8.4 Conclusion

This study has shown that black liquor recycled from NaOH pretreatment of EFB can be used as pretreatment solvent of sweet sorghum bagasse. The result showed performance similar to the pretreated SSB with 6% NaOH. The ethanol concentration from pretreated SSF by using black liquor at 30 min processing time was not significantly different with 60 min that reached 45–46 $g\ l^{-1}$. This research can be applied to other lignocellulosic materials with low lignin content, but further studies are needed.

Acknowledgments The authors thank Dr. Haznan Abimanyu for the discussion in this work. This research was funded by Unggulan Project of Indonesian Institute of Sciences (LIPI) of fiscal year 2015.

References

Amriani F, Barlianti V, Muryanto V, Sari AA (2015) Activated carbon from lignin-based black liquor coagulated by polyaluminium chloride. Procedia Chem 16:134–140

Bradshaw TC, Alizadeh H, Teymouri F, Balan V, Dale BE (2007) Ammonia fiber expansion pretreatment and enzymatic hydrolysis on two different growth stages of reed canarygrass. Appl Biochem Biotechnol 137–140:395–405

Cai H, Dunn JB, Wang Z, han J, Wang MQ (2013) Life cycle energy use green house gas emission of production of bioethanol from shorgum in United States. Biotechnol Biofuels 6:141

Cantarella M, Cantarella L, Gallifuoco A, Spera A, Alfani F (2004) Comparison of different detoxification method for steam-exploded poplar wood as substrate for the bioproduction of ethanol in SHF and SSF. Process Biochem 39:1533–1542

Cassman KG, Liska AJ (2007) Food and fuel for all: realistic or foolish? Bioprod Bioref 1:18–23

Chambergo FS, Borges FG, Andriani PP, de Oliviera GS, Valencia EY (2015) Biomass and bioenergy. In: Energy and science technology, Vol 7: Bioenergy. Studium Press LLC, Houston

Dahnum D, Tasum SO, Triwahyuni E, Nurdin M, Abimanyu H (2015) Comparison of SHF and SSF processes using enzyme and dry yeast for optimization of bioethanol production from empty fruit bunch. Energy Procedia 68:107–116

Dias MOS, Junqueira TL, Cavalett O, Cunha MP, Jesus CDF, Rossel CEV, Filho RM, Bonomi A (2012) Integrated versus stand-alone second generation ethanol production from sugarcane bagasse and thash. Bioresour Technol 103:152–161

Fan LT, Gharpuray MM, Lee YH (1987) Cellulose hydrolysis biotechnology monographs. Springer, Berlin, p 57

Fu K, Yue Q, Gao B, Sun Y, Zu L (2013) Preparation, characterization and application of lignin-based activated carbon from black liquor lignin by steam activation. Chem Eng J 228:1074–1082

Goshadrou A, Karimi K, Taherzadeh MJ (2011) Bioethanol production from sweet sorghum bagasse by Mucorhiemalis. Ind Crop Prod 34:1219–1225

Hahn-Hägerdal B, Jeppsson H, Skoog K, Prior BA (1994) Biochemistry and physiology of xylose fermentation by yeasts. Enzym Microb Technol 16(11):933–943

Harshvardhan K, Pal MK (2015) Biofuels: A step Towards green energy. In: Energy and science technology, Vol 7: Bioenergy. Studium Press LLC, Houston

Hendriks A, Zeeman G (2009) Review pretreatments to enhance the digestibility of lignocellulosic biomass. Bioresour Technol 100:10–18

Ingledew WM (1999) Alcohol production by Saccharomyces cerevisiae: a yeast primer, in the alcohol textbook, 3rd edn. Nottingham University Press, Nottingham

Ioelovich M (2015) Energetic potential of plant biomass as a renewable source of biofuels-a review. In: Energy and science technology, Vol 7: Bioenergy. Studium Press LLC, Houston

Ismael AA, Mohammad RR, Younis HA (2008) Utilizing calcium carbide for aqueous ethanol dehydration. Trikit J Pure Sci 13(1):176–178

Kang KE, Han M, Moon SK, Kang HW, Kim Y, Cha YL, Choi GW (2013) Optimization of alkali-extrusion pretreatment with twin-screw for bioethanol production from Miscanthus. Fuel 109:520–526

Li BZ, Balan V, Yuan YJ, Dale BE (2010) Process optimization to convert forage and sweet sorghum bagasse to ethanol based on ammonia fiber expansion (AFEX) pretreatment. Bioresour Technol 101:1285–1292

McKendry P (2002) Energy production from biomass (part 1): overview of biomass. Bioresour Technol 83:37–46

Millet MA, Baker AJ, Scatter LD (1976) Physical and chemical pretreatment for enhancing cellulose. Appl Microbiol Biotechnol 29:462–468

Mosier N, Wyman C, Dale B, Elander R, Lee YY, Holtzapple M, Ladisch M (2005) Features of promising technologies for pretreatment of lignocellulosic biomass. Bioresour Technol 96:673–686

Muryanto TE, Hendarsyah H, Abimanyu H (2015) Reuse black liquor of alkali preteatment in bioetanol production. Energy Procedia 68:236–243
Ohgren K, Bengtsson O, Gorwa-Grauslund ME, Galbe M, Hahn-Hagerdal B, zacchi G (2006) Simultaneous saccharification and co-fermentation of glucose and xylose in steam pretreated corn stover at high fiber content with Saccharomyces cerevisiae TMB3400. J Biotechnol 126 (4):488–498
Pronk JT, Steensmays HYDE, van Dijken JP (1996) Pyruvate Metabolism in Saccharomyces cerevisiae. Yeast 12:1607–1633
Rizk M, Antranikian G, Elleuche S (2012) End to end gene fusions and their impact on the production of multifunctional biomass degrading enzymes. Biochem Biophys Res Commun 428:1–5
Ruiz R, Ehrman T (1996) Chemical analysis & testing task, determination of carbohydrate in biomass by high performance liquid chromatography. Laboratory analytical procedure, NREL LAP-002
Saibi W, Brini F, Hanin M, Masmoudi K (2013) Development of energy plant and their potential to withstand various extreme environment. Recent Pat DNA Gene Seq 7:13–24
Sjostrom E (1993) Wood chemistry fundamentals and applications, 2nd edn. Academic, London
Sudiyani Y, Styarini D, Triwahyuni E, Sudiyarmanto SKC, Aristiawan Y, Abimanyu H, Han MH (2013) Utilization of biomass waste empty fruit bunch fiber of palm oil for bioethanol production using pilot – scale unit. Energy Procedia 32:31–38
Taherzadeh MJ, Karimi K (2008) Pretreatment of lignocellulosic waste to improve ethanol and biogas production: a review. Int J Mol Sci 9:1621–1651
Wan C, Zhou Y, Li Y (2011) Liquid hot water and alkaline pretreatment of soybean straw for improving cellulose digestibility. Bioresour Technol 102(10):6254–6259
Weimer PJ, Springer TL (2007) Fermentability of eastern gamagrass, big bluestem and sand bluestem grown across a wide variety of environments. Bioresour Technol 98:1615–1621
Wingren A, Galbe M, Zacchi G (2003) Techno-economic evaluation of producing ethanol from softwood. Comparison of SSF and SHF and identification of bottleneck. Biotechnol Prog 19 (4):1109–1117
Wu L, Arakane M, Ike M, Wada M, Takai T, Gau M, Tokuyasu K (2011) Low temperature alkali pretreatment for improving enzymatic digestibility of sweet sorghum bagasse for ethanol production. Bioresour Technol 102:4793–4799
Xu J, Zhang X, Jay JC (2012) Pretreatment of corn stover for sugar production with switchgrass-derived black liquor. Bioresour Technol 111:255–260
Zacchi G, Skoog K, Hagerdal H (1988) Economic evaluation of enzymatic hydrolysis of phenol-pretreated wheat straw. Biotechnol Bioeng 32(4):460–466

Chapter 9
Increasing Farmers' Affordability for Anaerobic Digester Construction in Rural Area Using Spatial Cluster Analysis

Christia Meidiana, Siska Ita Selvia, and Dian Dinanti

Abstract Spatial cluster analysis is used in this study to group non-biogas farmers since the unaffordability of the individual anaerobic digester (AD) is the main factor in slow biogas development in rural areas. Therefore, this study aims to determine the most reasonable manure biogas management in Argosari Village, Malang Regency, East Java Province. There are 234 farmers raising 596 cows and 44 farmer households who have been using manure waste as feedstock for AD. The biogas generated is mainly used for cooking. Before clustering, the individual affordability for the farmer is measured, while the group's affordability was estimated afterward under two conditions. The first condition is based on cow ownership, and the second is based on the AD capacity. GIS techniques were used to determine the cluster size and location. Cluster size is characterized by the number of households and the spatial proximity of the cluster members. Cluster location indicates the prospective AD location as it meets the maximum distance between the house and the AD. Before clustering, almost 100% of farmers cannot afford the AD since the individual cost burden is relative high. Clustering decreases the cost burden and increases the affordability consequently with two different percentages. AD size-based biogas management increases farmer's affordability more effectively than cow ownership-based biogas management does. The $4m^3$ and $6m^3$ AD are appropriate for the cluster, since most AD construction cost is supported by the local government. However, not all of the manure waste can be used as feedstock, allowing environmental pollution. Meanwhile, the cow ownership-based measurement increases the affordability but only for some farmers.

C. Meidiana (✉) • S.I. Selvia • D. Dinanti
Department of Urban and Regional Planning, Faculty of Engineering, Universitas Brawijaya, Jl. MT. Haryono 167, Malang 65147, Indonesia
e-mail: c_meideiana@ub.ac.id

B. McLellan (ed.), *Sustainable Future for Human Security*,
https://doi.org/10.1007/978-981-10-5430-3_9

9.1 Introduction

In Copenhagen COP 17 2009, the Government of Indonesia (GoI) announced its commitment to reduce greenhouse gas (GHG) emissions up to 26% by 2020. Currently, GHGs emissions are contributed by five leading sectors, i.e., energy, transportation, forestry, agriculture, and waste. Two sectors, energy and transportation, emit GHGs from fossil fuel combustion, while the other sectors produce GHGs from different sources. Fossil fuel is the main energy source in Indonesia, covering about 71% of the national energy consumption. This percentage is dominated by energy consumption in transportation and the energy sector causing high GHG emissions from these sectors which contribute 4.11% and 18.3%, respectively. GoI is working on reducing this percentage by encouraging regulation enforcement and compliance at the national and regional level supporting renewable energy (RE) development to increase RE utilization. For example, Presidential Regulation 5/2006 on National Policy on Energy regulating biogas as one of the REs which may be developed to attain national energy security. GoI aims to increase the share of RE in which 10.81% final energy consumption will have been from biomass including its derivatives (e.g., biogas) by 2020. Furthermore, new regulations on RE development have been endorsed to attain the target. The Ministry of Energy and Mineral Resource regulates RE utilization for electricity and feed-in tariff for biomass and biogas plant in Ministerial Regulation (MR) number 12/2017 and 27/2014, respectively. The latter legalizes the purchase of electricity from biomass and biogas power plants by PT PLN (a state-owned power utility company) and the increased feed-in tariff (FIT) for biomass/biogas generated electricity. Governmental Regulation 79/2014 focuses on National Policy on Energy targets which are 11% primary energy conservation as well as 17% and 23% increase of REs utilization by 2020 and 2025, respectively. In 2014, REs utilization was only 6% although the potential of RE in Indonesia is high. For example, it is estimated that there is potential for 32,000 MW energy from biomass (including its derivatives). However, the installed capacity is only 1740.40 MW and targeted to be 115 GW in 2025. Good implementation of the regulations may accelerate RE development and increase electrification ratio in 2014, which is 84.35% to 100% electrification ratio by 2020.

Biogas utilization has tended to increase in Indonesia because it offers many social, economic, and environmental benefits (Yuana and Zuo 2011). It increases the community awareness of green energy, reduces GHG emission, promotes sustainable development of the village, accelerates environmentally friendly agriculture, improves rural household savings, and increases rural energy equity as well as the quality of rural life (Chen 1997; Chen et al. 2010, 2011; Zhang et al. 2012). The final energy consumption shifts from conventional energy dependence to local renewable energy empowerment to save energy and to protect the environment (Venkateswara Rao et al. 2010; Yu et al. 2008). Though it has these benefits, the development of renewable energy in rural areas is relatively slow, due to the high initial cost which is mostly spent on production, research and development, and implementation (Anthony 2006; Klaas et al. 2008; Anna 2010; Kobos et al. 2006). Integrated rural biogas planning can contribute an appropriate solution to overcome the problem. However, biomass to biogas schemes are highly geographically dependent for two reasons. Firstly, biomass supply and biogas demand are often distributed dispersedly (Igoni et al. 2008). Secondly, there are some spatially related requirements, including technical requirements for determining the biogas plant locations, land suitability, availability, and proximity requiring a wide range of geographical data (Zubaryeva et al. 2012). Geographical information systems (GIS) approaches can accommodate this demand.

The GIS technique is widely used for analyzing spatially related issues as well as for assessing the potential biomasses for biogas production and for determining the location of a biogas plant in rural areas (Fiorese and Guariso 2010a; Kurka et al. 2012; Sultana and Kumar 2012; Sliz-Szkliniarz and Vogt 2012). Other authors (Hoehn and et al. 2014) applied a GIS approach to determine the biogas plant location by identifying the potential biomass sources. Evaluation of appropriate AD location based on the manure collection system was conducted by Dagnall et al. (2000) and Ma et al. (2005). Furthermore, a GIS-based tool was developed to calculate the available livestock manure and to estimate the available biogas by Batzias et al. (2005). GIS has also been studied for estimating the regional distribution of biogas potential based on livestock residues. GIS can be combined with other techniques such as multi-criteria analysis (MCA), including analytical hierarchy process (AHP) to plan bioenergy management, including biomass potential identification and appropriate AD location determination (Haddad and Anderson 2008; Fiorese and Guariso 2010b; Franco et al. 2015). In this paper, a combination between GIS techniques with cluster analysis is used to consider other aspects since the optimal location of biogas plants is more or less influenced by many aspects such as social and technical aspects, economic considerations, and environmental requirements. The absence of some important driving forces in AD location determination, i.e., social, technical, and economic aspect in the previous studies, has been accommodated in this study. Using cluster analysis, characteristics of different farmer households are identified and considered to calculate local acceptance rate, economic performance of bio-digesters, and appropriate AD capacity.

The economic performance of AD is important for farmers because AD is an investment and a net benefit is their target. The net benefit is determined by related aspects including affordability, subsidies, and cost load. In developing countries, most of the rural households cannot afford the price of the renewable energy, particularly biogas, since it is comparatively expensive due to high construction costs (Mulder and Tembe 2008; United Nations 2005). Authors (Qu et al. 2013) affirmed that the major barrier to biogas promotion is the investment on construction. The solution can be increasing affordability, providing suitable subsidy schemes or decreasing cost load. Efforts have been proposed through studies to reduce installation costs, i.e., by constructing low-cost digesters (Ferrer et al. 2011) or by providing effective and efficient subsidy schemes (Wang et al. 2016). It has been suggested (Culler and Webber 2009; Srinivasan 2008) that subsidies should be provided by the government, since biogas has lots of social and environmental benefits and (York et al. 2016) indicated that access to subsidies improved profitability. Meanwhile, others (Meidiana 2015) proposed a method to increase affordability by combining economic and spatial aspects. The result showed that constructing communal AD may increase farmer's affordability since they can share construction costs.

Recently, the regional government of East Java Province had a growing preference for decentralized systems especially individual household systems. This system better fits into the government's line of allocating finance aid for farmer households with various numbers of cow ownership. There were subsidies from the local government for farmers willing to construct an individual AD. The amount of the subsidy was not determined based on installed AD capacity depending on manure waste volume utilized. However, the subsidy was relative low compared to AD price so most of the price has to be covered by individual farmers. Preliminary surveys showed that few individual farmers have sufficient net income to afford AD. Furthermore, with an individual household system, monitoring of their correct operation is more feasible. This is the reason why communal systems are not commonly used in the region. The choice between individual and communal options for biogas systems lies at the core of concerns on environmental performance, economic performance, and social effects of infrastructures.

Therefore, this paper aims to measure the non-biogas farmer's affordability of anaerobic digester (AD) in the area of study because the village is targeted to be a self-sufficient energy village (SSEV) by 2020. As a SSEV, Argosari Village should cover energy demand sufficiently with 60% of energy coming from renewable energy. Since the village has potential of RE source from manure waste, it is important to maintain the sustainability of biogas development to achieve the target. Preliminary surveys indicated that non-biogas farmers in Argosari Village generally have low affordability to construct the individual AD causing slow biogas development. This study contributes to a more evidence-based spatial planning on measuring farmer's affordability in rural area by comparing the advantages and disadvantages of both decentralized biogas system and concluding under which conditions each should be intended. GIS is used as a supporting tool in determining AD location since the village is located in a relatively hilly area characterized by contour landscape requiring well-planned biogas development. To do so, the next section outlines the material and methods, describing the village characteristics, biogas source, and methods used in the study. Subsequently, the analysis is presented and discussed with respect to economic performance and energy production for both decentralized system. Lastly, conclusions are formulated and some points are recommended for better future studies.

9.2 Materials and Methods

9.2.1 Area of the Study

Argosari Village was selected as a pilot project for biogas development program in Malang Regency in 2005 because it has livestock potential. The geographical setting which is a hilly landscape and the mild climate condition in Argosari Village are appropriate for the developing the agricultural sector, especially husbandry. The village is located about 32 km east of Malang City and covers an area of 421,423 ha comprising three smaller units called dusun. Farming is the main livelihood in the village where 596 cows are raised by 234 farmers, producing about 2080 kg manure waste per day (Table 9.1). The village won the prize for green energy village in 2011 since some farmers have utilized manure waste for the biogas production process. However, the number of anaerobic digesters decreased as many of them were damaged. Initially, there were 34 ADs constructed as pilot projects financed fully by the local government. In the next 3 years, the number of AD increased although the financial support changed from fully to partly supported (subsidy) indicating higher farmer's acceptance to biogas technology. The subsidy was provided for individual AD and was a fixed amount independent from the AD capacity. This means that the higher the capacity is, the more cost load has to be born. In 2012, the number of AD decreased from 261 to 121 or more than 50% decline which continued in 2014. Currently, there are 44 AD showing 5% biogas utilization because only 44 out of 234 farmers have used manure waste as feedstock for AD (Fig. 9.2). Some farmers spread fresh manure over the field for fertilizer, but most farmers dispose manure waste to the ditches or streams leading to water pollution and odor. The anaerobic digestion process in Argosari Village is illustrated in Fig. 9.3 where fresh manure derived from cow stables is collected manually. The manure is extracted from the dung pit and then fed into AD. All biogas farmers (44 farmers) use fixed dome-type AD with the various capacities ranging between 4 m^3 and 8 m^3.

Argosari Village borders the national park of Bromo Tengger-Semeru. Therefore, some households search for wood in the national park for cooking. The number of incidence of illegal tree cutting by the villagers has been increasing since 2012 along with the increasing fuel price. Therefore, the local government has been promoting the biogas management in the village to decrease the number of illegal logging cases.

Table 9.1 Farmers in Argosari Village

No	Dusun	Number of		
		Biogas farmer	Non-biogas farmer	Total
1	Pateguhan	–	20	20
2	Gentong	2	11	13
3	Bendrong	42	159	201
	Total			234

Table 9.2 Main categories of the questions in the survey

Category	Questions
Household	Name, education, occupation, number of members in family, land area, income, expenses
Animal production	Animal category (dairy or beef cattle), number of animals, weights of animals, feeding methods
Manure management	Frequency of cleaning of animal house (times/day)
	Volume of water used (m^3 day/1)
	Distance from stall to house and drinking
	Water resource
	Manure collection system (manual or mechanical)
	End use of manure (crop, fish pond, biogas, discharge to waterways)

9.2.2 Data Collection and Analysis

A survey set out to investigate the current situation of biogas utilization and manure management in the village. All the obtained data were rechecked to ensure that the data were correct. After the checking of the data, the final survey included information from all samples (194 non-biogas farmers). The data collected included the socioeconomic background of the household, manure waste characteristic of the farms, and technical aspects. The questionnaire for this study was prepared through workshops involving experts in manure management, rural planning, and RE management. Questions were formatted focusing on the following three main areas: information about the household, animal production, and manure management (Table 9.2).

9.2.3 Description of the Manure Waste

The average manure generation in Argosari Village is 29.8 kg/head/day. Raw cow manure is mixed with some amount of water to get a slurry concentration enabling stirring and flowing and the slurry flows to fixed-dome AD. The biogas production rate of each household (HH) depends on cow ownership ranging from 2 to 4 head/HH. The organic dry matter (ODM) content is roughly 83%, and it is assumed that adding water decreases the dry matter content to 9.5% since none of biogas farmers measure the water amount. This value is adopted from the previous related study (Qu et al. 2013). Biogas production from AD biogas farmers was observed as being 276 m^3/t ODM. The salient features of substrates are presented in Table 9.3 and are used for energy generation calculation.

9.2.4 Affordability

One of the main factors of rural biogas development is the affordability describing the individual ability to purchase goods or services. The term "affordability" has

Table 9.3 Manure waste description

Parameter	Unit of measurement	Value
Average manure production	kg/d	29.8
Total manure	kg/d	22,080
Input manure for existing AD	kg/d	5070
Dry matter	%	9.5
Organic dry matter (ODM)	%	83
Biogas yield	m3/t ODM	276
Methane concentration	%	55

different contexts depending on the way it is discussed. For example, authors (Winkler et al. 2011) cite the Energy Sector Management Assistance Program, 2003; "Affordability is a politicized concept," while others (Riley 2014) define affordability as the net change in income – expenditure. This paper defines affordability in a specific mathematical way according to (Yudariansyah et al. 2006). Affordability (A) is the ratio of the net income to the good's price (P) where net income is calculated by subtracting expenses (E) from gross income (Ig) in certain periods (see Eq. 9.1):

$$A = (Ig - E)/P \qquad (9.1)$$

The AD is considered affordable for the farmer when the ratio is greater or equal to one. The affordability is compared under three conditions which are individual, cow ownership-based group, and AD capacity-based group. The affordability change under these conditions is measured to determine the appropriate biogas management in Argosari Village.

9.2.5 *Spatial Cluster Analysis*

Spatial cluster analysis is a geographical quantitative analysis used to determine the distribution pattern of the data of objects based in proximity describing the similarity of the objects. The proximity is measured using K-nearest neighbor method including the boundary of the cluster indicated by mean distance. Spatial cluster analysis is used to determine the mean distance which is appropriately applied then as an average maximum radius in which some objects can be grouped into the same cluster. These objects will be considered as a group or cluster if the data meet the critical value limit. The value should be less than 1, less than 2.58, and less than 0.01 for nearest neighbor ratio, z-score, and p-value, respectively.

9.2.6 Energy Calculation

Since Argosari Village is targeted as an SSEV, it is essential to calculate the potential energy production from manure biogas in order to predict whether the target can be achieved or not in the future. In this study, the energy calculation is conducted to measure the energy potential from the anaerobic digestion process and to compare the amount of energy generated from individual AD and communal AD. As described in Fig. 9.1, all farmers in Argosari Village use biogas for cooking. Therefore, the heat energy generation from individual AD and communal AD is calculated using Equation (Eq. 9.2) (Akbulut 2012)

$$E_{\text{th}} = \frac{m_{\text{manure}} \times \text{DM}_{\text{manure}} \times \text{ODM}_{\text{manure}} \times \text{BY}_{\text{manure}} \times 5.5 \times \eta_{\text{th}}}{t} \quad (9.2)$$

where Eth (kWh/h), 5.5 (kWh/m^3), BY (m^3/tODM), and ηel (45%) are the amount of heat energy from biogas, the total energy value of biogas, conversion index, and

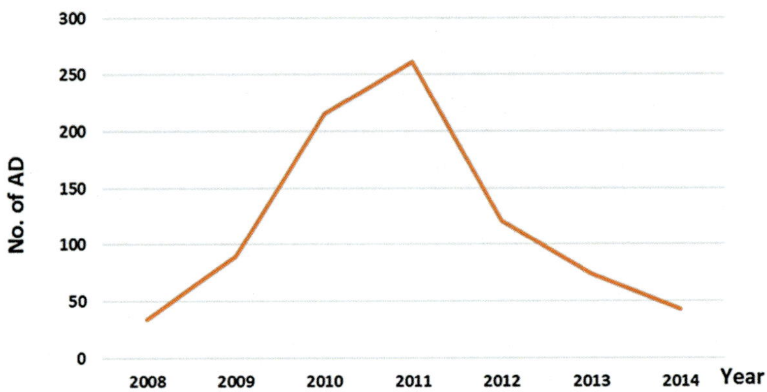

Fig. 9.1 AD trend in Argosari Village

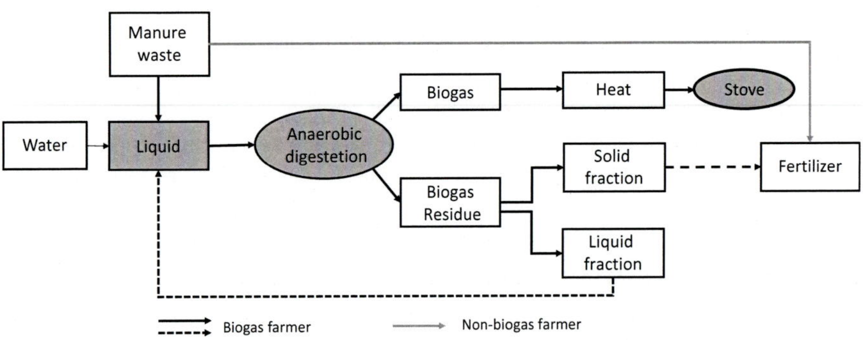

Fig. 9.2 Biogas production system in Argosari Village

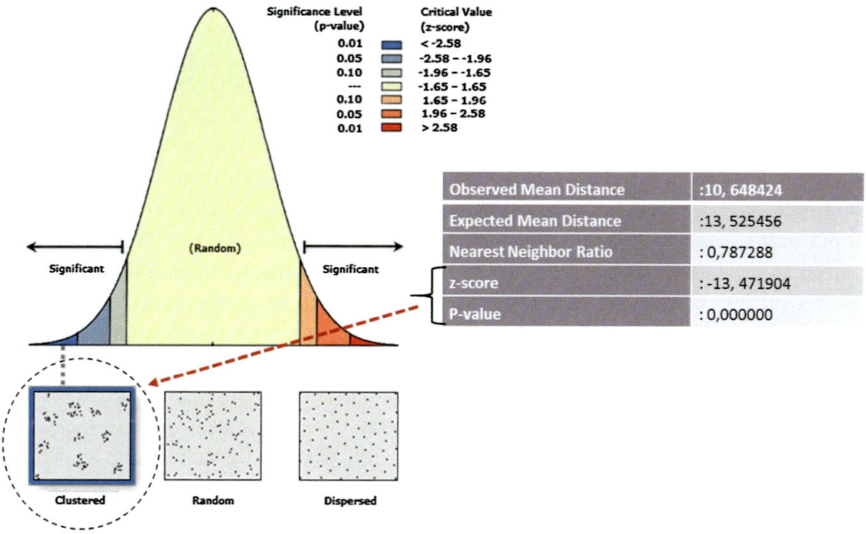

Fig. 9.3 Spatial cluster analysis result

thermal efficiency value, respectively. Operating time parameter, indicating the length of biogas processing, t (h/year) varies typically from 8000 to 8760. The minimum operating time parameter applied in this study is 8000 (h/year).

9.2.7 Assumptions and Limitations

Some assumptions were used in this paper to enable better analysis. Assumptions were made based on related previous studies, local current conditions, and available data which are:

1. Households raise their dairy cattle as assets and keep them for a relatively long time. Since the target of biogas management is SSEV, it is important to maintain the number of cattle. Based on the survey, most farmers are raising the cattle from young (calf) until their unproductive age.
2. All cows have the same age (productive age) with the average weight 400–500 kg and generate 25 kg–30 kg manure per day. Average manure waste was calculated based on the average manure production of each household.
3. Manure waste characteristic is generalized for the whole cows since there is no field measurement for manure samples.

9.3 Results and Discussion

The two-month survey was conducted to collect the data of socioeconomic and demographic household (HH) in Argosari Village which is described in Table 9.4. AD with the capacity of 6 m^3, 8 m^3, 10 m^3, and 12 m^3 is the common type in this village to process the manure waste. The AD price is proportional to the capacity ranging between Rp. 4,500,000 and Rp. 12,000,000 (see Table 9.5). As mentioned in the previous section, the number of AD decreased from 2011 until 2014, and it started to increase in 2016 with only one new AD constructed. From the survey, it was found that former biogas farmers and the non-biogas farmers have interest in constructing AD with the local acceptance being 100%. However, there are four factors hindering biogas utilization in the village which are finance shortage, labor unavailability, lack of information, cow insufficiency, and land inadequacy with the percentage of 47%, 27%, 15%, 7%, and 4%, respectively. It obvious that cost load is the main obstacle for slow biogas development.

Based on the cow ownership, the affordability was calculated. From Table 9.6, it is obvious that the number of the farmers who cannot afford the construction cost of individual AD is relatively high. Low household gross income which lies between approximately 500,000 and 1,000,000/month is the reason for this. These farmers can only cover 20–70% of the total cost without considering their monthly expenses. It indicates that the cost load of individual farmer households to construct AD is well beyond their net income. Therefore, clustering is applied grouping individual farmers to increase affordability.

Table 9.4 Socioeconomic and demographic household data from survey results

Parameter	Unit of measurement	Value
Average household size	Number of persons	4.1
Average cattle herd size owned	Number of animals	3.4
Average cooking energy demand (LPG)	Kg/hh	6
Average number of cooking times/household/day	Frequency	2.6
Monthly income	Rp/hh	300,000–5,000,000
Monthly LPG cost for cooking purposes	Rp/hh	35,000
Monthly savings	Rp/hh	25,000–800,000

Table 9.5 Size of anaerobic digester in Argosari Village

AD size (m^3)	Required number of cow (head)	Manure mass for feedstock (Kg)	Price (Rp)
4	3–4	20–40	4,500,000
6	5–6	40–60	6,000,000
8	7–8	60–80	8,000,000
10	9–10	80–100	11,000,000
12	11–12	100–120	12,000,000

Table 9.6 The individual HH's affordability of AD construction cost

Dusun	Number of HH and suitable AD size based on the cow ownership					Affordable	Not affordable
	4 m^3	6 m^3	8 m^3	10 m^3	12 m^3		
Pateguhan	3	1	0	4	4	11%	89%
Gentong	3	4	0	0	2	15%	85%
Bendrong	1	1	0	4	12	9%	91%
Total	7	6	0	8	18		

Table 9.7 Parameter value of in spatial cluster analysis

Parameters	Criteria	Value
Observed mean distance	–	10.648424
Expected mean distance	–	13.525456
Nearest neighbor ratio	< 1	0.787288
Z-score	< 2.38	−13.471904
P-value	< 0.01	0.000001

Using the spatial cluster analysis, HHs are grouped into clusters. The parameter values for clustering are presented in Table 9.7. All values meet the criteria for clustering, which are 0.78, −13.471, and 0.00 for nearest neighbor ratio, z-score, and p-value, respectively. These values showed that the settlement form in Argosari Village is clustered. This pattern enables communal AD construction consisting of 2–5 households. The observed and expected mean distance is 10.65 meters and 13.53 meters, respectively (Fig. 9.3). The cluster is formed with a maximum radius of 11 meters instead of 14 meters considering that shorter distance includes fewer houses to be one cluster enabling easier biogas management due to technical requirements (proximity to the stall, slope, and distance between houses).

Based on the radius, there are a total of 32 clusters in Argosari Village. One AD may be equipped to each cluster for processing manure waste. In each cluster, there are 1–4 households (HHs) as described in Table 9.8. Most clusters are located in Dusun Bendrong since most non-biogas farmers live in Dusun Bendrong. The distance between houses is relative shorter compared to other dusuns. Geographical settings also affect the pattern since it is located in the highlands.

The calculation comes to the result that affordability increases after clustering based on the cow ownership (Table 9.9). Yet, the affordability rate in Argosari Village is relatively low if manure from all cows in the group is used for feedstock. Not all farmers can afford the construction cost through the clustering. The proportional price to AD capacity is the reason for this condition. The AD with maximum capacity is more expensive compared to small and medium capacity AD (4m^3 and 6m^3).

Therefore, the cluster conditions are set where AD capacity is set first as 4 m^3 and 6m^3 to determine the number of the cows whose manure can be used for feedstock.

Table 9.8 The number of cluster and its members

Dusun	Number of HH in one cluster		
	1 HH	2 HH	3 HH ≤
Pateguhan	2	4	2
Gentong	3	7	1
Bendrong	6	5	2

Table 9.9 Affordability of the farmers after clustering based on the cow ownership

Dusun	No. of cluster	No. of HH in each cluster	No. of cows used/HH	AD size [m^3]	Cost burden [Rp]	Affordability
Pateguhan	6	1–5	1–6	4–12	750,000–4,200,000	6%–35%
Gentong	4	1–4	1–4	4–12	750,000–4,425,000	6%–37%
Bendrong	9	1–6	1–7	4–12	750,000–6,350,000	7%–53%

Table 9.10 Affordability of the farmers after clustering based on the AD size

Dusun	No. of cluster	No. of HH in each cluster	No. of cows used/HH	AD size [m3]	Cost burden [Rp]	Affordability
Pateguhan	6	1–5	1–6	4 & 6	750,000–4,200,000	13%–70%
Gentong	4	1–4	1–4	4 & 6	750,000–4,425,000	13%–74%
Bendrong	9	1–6	1–7	4 & 6	750,000–6,350,000	13%–106%

Table 9.11 The energy production from clusters based on two different conditions

Dusun	No. of cluster	No. of HH in each cluster	Energy generation [kW/a]	
			Cow ownership	AD size
Pateguhan	6	1–5	1.50	0.67
Gentong	4	1–4	0.97	0.67
Bendrong	9	1–6	2.77	1.35
Total			5.24	2.69

The results show that the affordability rate increases compared to the previous condition where cow ownership is used to determine AD size (Table 9.10). By forming a group, affordability can be increased significantly, particularly for clusters consisting of three or more households, because of more cows generating more manure waste.

The energy production is dependent on the size of the AD. Therefore, clusters with maximum cow manure utilization produce more energy compared to cluster with optimum cow manure utilization (Table 9.11). According to cluster analysis results, Argosari Village can form clusters with AD capacity of 10–12 m^3 enabling high-energy generation which can be distributed to the members of the cluster. Thus, the villagers gain more environmental benefits in the first condition because untreated manure waste can be minimized significantly especially in Dusun Bendrong where most cattle are raised.

9.4 Conclusion

Spatial cluster analysis was used in this study to measure the farmer affordability of AD construction cost. The analysis is conducted by means of GIS. Generally, GIS offers flexibility to include different spatial and nonspatial data (income, expenses, construction cost, livelihood, and education) that quantify the techno-economic biogas potential. Spatial data were measured to identify whether objects can be clustered or not and to calculate the mean distance to determine the cluster size. The mean distance indicates the radius of the cluster and determines the number of households (HHs) in each cluster as members of the cluster. Most non-biogas farmer HHs cannot afford the AD construction cost before clustering. There are only 28 HHs (11.7%) that can afford the cost, while 206 HHs (88.3%) cannot. The affordability after clustering is higher than before clustering. The affordability increases more when biogas management is determined based on the optimum size of AD instead of the cow ownership. The affordability of cow ownership-based biogas management and AD capacity-based biogas management ranges between 6%–53% and 13%–106%, respectively. It can be concluded that the affordability is influenced by the number of HHs involved in AD construction and the number of cows whose manure is used for feedstock. Also, since the energy production depends on the amount of the manure waste, the first condition generates more energy compared to the second one amounting to 5.24 kW/year and 2.69 kW/year, respectively.

9.5 Recommendations

Local governments may take benefits from the study since the method is applicable for other villages and even other regions as long as all variables required for data analysis are available. Adequate data, i.e., village map, number of households, cattle ownership, and energy consumption pattern, are required for good results. In the case of Argosari Village, the data is relatively complete. However, some assumptions had to be made for missing data which does not affect the results significantly. For future-related studies, these assumptions should be eliminated for precise results based on updated local conditions. Furthermore, future studies should determine the AD location for each cluster by considering geographical settings and technical aspects. AD location may affect the construction cost that must be carried by the members because it determines feeding and distribution pipe length from and to each household within the cluster.

Based on the conclusion, the following points should be implemented based on the findings to increase the biogas utilization rate in Argosari Village which are:

1. Appropriate subsidy scheme for AD construction since the current scheme does not consider about the cost load proportional to AD capacity. Furthermore, the

scheme for communal AD may be provided to accommodate farmer group with one AD since the result affirms that clustering increases farmer's affordability.
2. Local government may offer biogas farmers incentives for distributing biogas to their neighbors outside the cluster (non-farmer households). This can motivate biogas farmers since they know that side income can be earned from monthly distribution fee. Proximity is considered as priority to determine the beneficiaries. The remaining biogas has to be calculated beforehand to estimate the number of beneficiaries
3. Non-biogas farmers have no possibility to be a member of cluster because of maximum distance requirement and insufficient cows to construct individual AD can commend their cattle to the nearest cluster.
4. Existing farmer groups can be empowered to disseminate biogas technology among the members. Biogas farmers and local governmental staffs may be involved as information sources.
5. Capacity building to improve the skills of the farmers including better cattle management, proper AD construction, and biogas management should be provided by the local government.
6. A revolving fund may be introduced by the local government to maintain the biogas sustainability. Capable nongovernmental organizations (NGO), i.e., Dutch NGO, HIVOS which have much experiences in rural biogas development in East Java Province, can be involved to manage the fund.

References

Akbulut A (2012) Techno-economic analysis of electricity and heat generation from farm-scale biogas plant: Çiçekdagı case study. Energy Energy 44:381–390

Anna P (2010) Renewable energy in South Africa: potentials, barriers and options for support. Energy Policy 38(9):4945–4954

Anthony DO (2006) Renewable energy: externality costs as market barriers. Energy Policy 34 (5):632–642

Batzias FA, Sidiras DK, Spyrou EK (2005) Evaluating livestock manures for biogas production: a GIS based method. Renew Energy 30:1161–1176

Chen R (1997) Livestock-biogas-fruit systems in South China. Ecol Eng 8:19–29

Chen Y, Yang G, Sweeney S, Feng Y (2010) Household biogas use in rural China: a study of opportunities and constraints. Renew Sustain Energy Rev 14:545–549

Cheng S, Li Z, Shih J, Du X, Xing J (2011) A field study on acceptability of 4-in-1 biogas systems in Liaoning province. China Energy Proc 8:1382–1387

Culler, A.D., Webber, M.E. 2009 Policy incentives, barriers and recommendations for biogas production. In: Proceedings of the ASME 2009. 3rd International Conference of Energy Sustainability. SanFrancisco, pp 1–8

Dagnall S, Hill J, Pegg D (2000) Resource mapping and analysis of farm livestock manures e assessing the opportunities for biomass-to-energy schemes. Bioresour Technol 71:225–234

Ferrer I, Garfıet M, Uggetti E, Ferrer-Martí L, Calderon A, Velo E (2011) Biogas production in low-cost household digesters at the Peruvian Andes. Biomass Bioenergy 35:1668–1674

Fiorese G, Guariso G (2010a) A GIS-based approach to evaluate biomass potential from energy crops at regional scale. Environ Modell Softw 25:702–711

Fiorese G, Guariso G (2010b) A GIS-based approach to evaluate biomass potential from energy crops at regional scale. Environ Model Softw 25:702–711
Franco C, Bojesen M, Hougaard JL, Nielsen K (2015) A fuzzy approach to a multiple criteria and geographical information system for decision support on suitable locations for biogas plants. Appl Energy 140:304–315
Haddad MA, Anderson PF (2008) A GIS methodology to identify potential corn stove col-lection locations. Biomass Bioenergy 32:1097–1108
Hoehn et al (2014) A Geographical Information System (GIS) based methodology for de-termination of potential biomasses and sites for biogas plants in southern Finland. Appl Energy 113:1–10
Igoni AH, Ayotamuno MJ, Eze CL, Ogaji SOT, Probert SD (2008) Designs of anaerobic di-gesters for producing biogas from municipal solid-waste. Appl Energy 85(6):430–438
Klaas VA, Huden SK, Marko PH (2008) Policy measures to promote the widespread utilization of renewable energy technologies for electricity generation in the Maldives. Renew Sust Energ Rev 12(7):1959–1973
Kobos PH, Erickson JD, Drennen TE (2006) Technological learning and renewable energy costs: implications for US renewable energy policy. Energy Policy 34(13):1645–1658
Kurka T, Jefferies C, Blackwood D (2012) GIS-based location suitability of decentralized, medium scale bioenergy developments to estimate transport CO2 emissions and cost. Biomass Bioenergy 46:366–379
Ma JG, Scott NR, DeGloria SD, Lembo AJ (2005) Siting analysis of farm-based centralized anaerobic digester systems for distributed generation using GIS. Biomass Bioenergy 28:591–600
Meidiana C, Rafsanjani A (2015) The Spatial-economic Approach for Determining Bio-gas Management in Rural Area. Int J Appl Eng Res 10(95):31–35
Mulder P, Tembe J (2008) Rural electrification in an imperfect world: a case study from Mozambique. Energy Policy 36:2785–2794
Qu W, Tu Q, Bluemling B (2013) Which factors are effective for farmers' biogas use?–evidence from a large-scale survey in China. Energy Policy 63:26–33
Riley PH (2014) Affordability for sustainable energy development products. Appl Energy 132:308–316
Sliz-Szkliniarz B, Vogt J (2012) A GIS-based approach for evaluating the potential of bio-gas production from livestock manure and crops at a regional scale: a case study for the Kujawsko-Pomorskie Voivodeship. Renew Sustain Energy Rev 16:752–763
Srinivasan S (2008) Positive externalities of domestic biogas initiatives: implications for financ-ing. Renew Sustain Energy Rev 12:1467–1484
Sultana A, Kumar A (2012) Optimal siting and size of bioenergy facilities using geographic information system. Appl Energy 94:193–201
United Nations. Increasing global renewable energy market share: recent trends and perspectives. United Nations (UN) Foundation, Beijing International Renewable Conference 2005
Venkateswara Rao P, Baral SS, Dey R, Mutnuri S (2010) Biogas generation potential by anaerobic digestion for sustainable energy development in India. Renew Sust Energ Rev 14:2086–2094
Wang C, Zhang L, Zhang Y, Pang M (2016) Alternative policies to subsidize rural house-hold biogas digesters. Energy Policy 93:187–195
Winkler H, Simoes AF, La Rovere EL, Alam M, Rahman MA, Mwakasonda S (2011) Access and affordability of electricity in developing countries. World Dev 39(6):1037–1050
York L, Heffernan C, Rymer C (2016) The role of subsidy in ensuring the sustainability of small-scale anaerobic digesters in Odisha, India. Renew Energy 96:1111–1118. ISSN 0960-1481
Yu L, Yaoqiu K, Ningsheng H, Zhifeng W, Lianzhong X (2008) Popularizing household-scale biogas digesters for rural sustainable energy development and greenhouse gas mitigation. Renew Energy 33:2027–2035
Yuana X-l, Zuo J (2011) Pricing and affordability of renewable energy in China – a case study of Shandong Province. Renew Energy 36(3):1111–1117

Yudariansyah, Had. et al.(2006) Analisis Keterjangkauan daya beli masyarakat terhadap ta-rif air bersih (PDAM) Kota Malang (Studi Kasus Perumahan Sawojajar)
Zhang LX, Wang CB, Song B (2012) Carbon emission reduction potential of a typical household biogas system in rural China. J Clean Prod 47:415–421
Zubaryeva A, Zaccarelli N, Del Giudice C, Zurlini G (2012) Spatially explicit assessment of local biomass availability for distributed biogas production via anaerobic co-digestion – Mediterranean case study. Renew Energy 39:261–270

Part IV
Marine Ecosystems

Chapter 10
Marine Ecosystem Sustainability Post-Mine Closure Activities (Macrobenthos Dynamics Study Due to the Influence of Tailing Disposal in Buyat Bay, Minahasa, Indonesia)

Djoko Hartoyo, A. Harsono Soepardjo, Abimanyu T. Alamsyah, and Arie Herlambang

Abstract This chapter explores the dynamics of macrobenthos that are affected by the disposal of tailings in Buyat Bay, Minahasa, Indonesia. Disposal of tailings on the seabed (submarine disposal) in Buyat Bay is is influenced by dispersion in water, which impacts the marine environment and ecosystem. This marine ecosystem is affected by the ocean conditions. Ideally, the condition of the sea with tailings disposal and land filling is such that the deep sea and ocean dynamics are not negatively affected and that the tailings do not harm the marine ecosystems. This study aims to evaluate the dynamics of marine ecosystems after the disposal of tailings in Buyat Bay. Previous incidental and temporal studies have been performed. However, studies that holistically assess the impact of tailings disposal on marine ecosystems with long-term time series data covering a time span of 10 years after the disposal have not been performed. This study is important because the sustainability of marine ecosystems will ultimately affect human security. Sediment samples were analyzed in laboratories. This study shows the marine ecosystem dynamics based on the diversity of marine biota (macrobenthos) over 10 years and illustrates the stages of succession in the marine ecosystem following tailing disposal as well as indicates the linkage between the environment approaches microbenthos take to maintain their viability.

D. Hartoyo (✉)
Environmental Science, University of Indonesia, Jakarta 10430, Indonesia
e-mail: djokohartoyo@gmail.com

A.H. Soepardjo
Physics Department, Faculty of Mathematics and Natural Sciences, University of Indonesia, Depok 16424, Indonesia

A.T. Alamsyah
Faculty of Technic, University of Indonesia, UI Campus, Depok, West Java 16424, Indonesia

A. Herlambang
Agency for the Assessment and Application of Technology, Jakarta, Indonesia

B. McLellan (ed.), *Sustainable Future for Human Security*,
https://doi.org/10.1007/978-981-10-5430-3_10

10.1 Introduction

Mining activities throughout the world, especially for gold, are very important to the economy. Although environmental issues always arise as a result of gold mines that use mercury and arsenic, which contaminate the soil, new gold mines continue to emerge (Salomons 1995). Technological innovations in the processing of gold ore include cyanidation in gold extraction to replace mercury. Furthermore, residual tailings from gold mines began to be dumped onto the seafloor in 1971, initiated by Canada. Several other countries have used a similar process for gold mining, such as in Indonesia in the Newmont Minahasa Raya (NMR) in Minahasa North Sulawesi and Newmont Batu Hijau in Sumbawa, West Nusa Tenggara.

The gold mining (PT NMR) activities in North Sulawesi Province began operations in March 1996 and ended on 31 August 2004. PT NMR utilized the seabed to dispose tailing waste generated from the mining process at a depth of 82 meters at a distance of about 900 meters from the coastline through 20 cms diameter of pipes. According to PT NMR, the selection of the submarine tailing placement system was based on environmental, engineering, and economic factor assessment.

PT NMR discharged about 2000 tons of tailings each day into Buyat Bay. Tailings slurry, as a by-product of mining activities, is designed to have a weight percent range that is between 45% and 55% solid and clay fractions with a density of about 1.336 kg/m^3 (Edinger et al. 2007). Compared with the density of seawater, 1.028 kg/m^3, the tailings are expected to settle to the seabed such that they are not dispersed into the sea (Ministry of Environment 2000). In fact, the sediments of tailings spread to a distance of 1–2 km from the discharge pipe (Edinger et al. 2007).

Theoretically the assumptions and the technology used may be true, but what happens at the PT NMR site shows the degree of accuracy of assumptions and technologies to be in doubt. Several times the tailings pipes were broken. Moreover, a thermocline that is relied upon as an STD safety layer was considered to be present based on the evidence of the mining company. This layer is considered to prevent the tailing lift-up to the mixed layer. However, the study conducted by PT NMR was still considered doubtful by many parties. Turbulence and upwelling factors could spread the tailings to the environment.

In Indonesia until now, there is no provision that regulates the ban on submarine tailing disposals. At the international level, there is also no regulation on tailings disposal on the seabed (UNCLOS 1982). As long as it meets the requirements and gets permission from the State Minister of the Environment, the placement of tailings on the seabed can be undertaken. The main requirement is to have an environmental impact analysis (AMDAL) in accordance with the provisions of the prevailing laws, namely, Law Number 23 of 2007 on Environmental Management and Law No. 32 of 2009 on the Protection and Management of the Environment.

Based on the disposal of tailings on the seabed (Sub Marine Disposal) in Buyat Bay, according to water dispersion, the tailings were predicted to affect the marine environment and ecosystems due to the influence of ocean dynamics on the condition of these tailings. The seabed conditions are ideal for tailings disposal and land filling in the deep sea because they are not affected by the dynamics of the ocean and are isolated; therefore, it is thought that the tailings do not impact the ecosystem (Ellis and Ellis 1994; Jones and Ellis 1995; Poling 2002; Ellis 2008).

The range of thermocline layers in tropical waters is at depths of 75 to 250 meters. This thermocline layer occurs throughout the year in tropical waters, whereas in temperate regions, it only occurs in summer. In the layer below the thermocline, these tailings are not expected to be lifted into the surface mixing layer above due to different water densities.

Previous research results recommended tailings disposal at Newmont Minahasa Raya below 135 meters depth (Prisetiahadi and Yanagi 2008). As a result, disposal of tailings on the seabed was performed for the Newmont Minahasa Raya gold mine in North Sulawesi, and this study seeks to examine its impacts. It was thought that this disposal would affect the marine environment and ecosystems as well as impact the communities around Buyat Bay due to changes in the marine ecosystem. Increases in suspended solids in the seawaters will increase its turbidity, decreasing the rate of phytoplankton photosynthesis. As a result, the primary productivity of water would be lowered, disrupting the entire food chain. This study is therefore important because the sustainability of marine ecosystems will ultimately affect human survival.

10.2 Materials and Methods

10.2.1 Area Studied and Data Collected

The study area was conducted in Buyat Bay, South Minahasa, North Sulawesi. Data collection from the primary data field in Buyat Bay occurred in the period from 2004 to 2014. This phase is intended to complement the time series data (Fig. 10.1).

Samples were taken at points from a predetermined station. Macrobenthos samples were collected from nine locations in Buyat Bay: two locations are at the tailings in the river mouth, one location is in the peak tailings site, four locations are around the tailings, and two locations are controls. The point sampling locations are shown in Fig. 10.2.

10.2.2 Analysis Methods

Samples were collected at the predetermined locations. Sediment samples were collected using the grab approach. The environmental quality was obtained from analyzing samples in the laboratory. Macrobenthos data were obtained and analyzed in the laboratory; the number and types or organisms were calculated and then analyzed by calculating the relative abundance (RA), diversity index (H'), uniformity index, (e) and dominance index (C). Time series analysis is used to describe

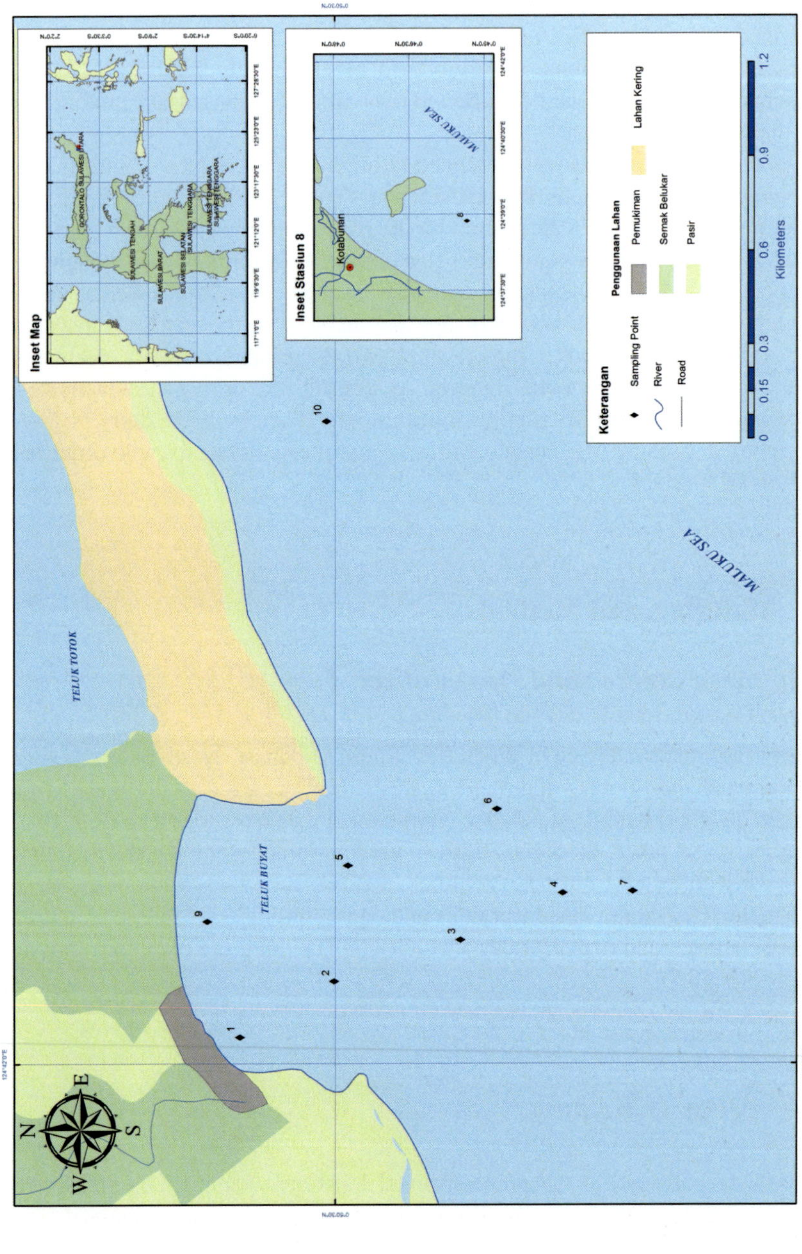

Fig. 10.1 Map of Buyat Bay, South Minahasa, North Sulawesi

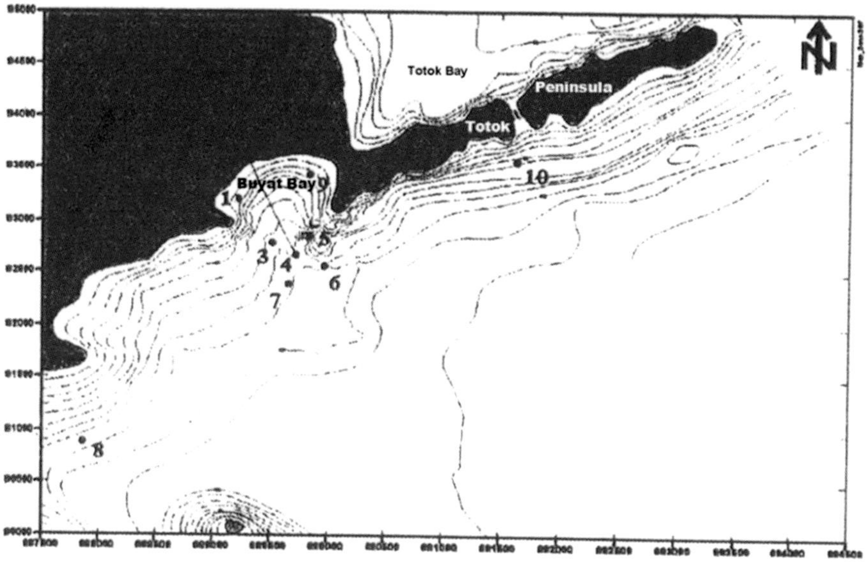

Fig. 10.2 Point sampling locations

the variability of the data series after 10 years for the dominant periodic functions (Emery and Thomson 2001).

RA, H', E, and C were evaluated using the following formula:

The relative abundance was measured using the following formula (Odum 1993):

$$\mathrm{RA} = 100\%(n_{\mathrm{i}}/N)$$

where:

RA = relative abundance (%)
n_{i} = one particular species found
N = total number of individuals found

The diversity index was measured using the formula for the Shannon-Wiener diversity index (Krebs 1989):

$$H' = -\sum_{i=1}^{s} \rho_i \log \rho_i$$

where:

H' = Shannon diversity index
p_{i} = proportion (n/N) of individuals
n_{i} = one particular species found
N = total number of individuals found

The uniformity index was measured using a formula according to Krebs (Krebs 1989):

$$e = H/\ln S$$

where:

e = uniformity index
H = diversity index
S = number of species

The dominance index was measured using the formula for Simpson's dominance index (Odum 1993):

$$C = \sum \left(\frac{n_{\mathrm{i}}}{N}\right)^2$$

where:

C = dominance index
n_{i} = one particular species found
N = total number of individual found

10.3 Results and Discussion

This study reveals the occurrence of marine ecosystem dynamics as shown by the value of the diversity of marine biota (macrobenthos) based on 10 years of monitoring. At the beginning of the study, the community structure showed a disturbance of benthic communities, which may reduce diversity. The Shannon-Wiener diversity index of benthos was lowest in Buyat Bay, especially in the area of tailings accumulation. The Shannon-Wiener diversity index in this area is at a value below 1, indicating that the area was negatively affected by severe pollution. Disturbances in this region are large over a period of approximately 8 years (1996–2004).

The absence of a thermocline with the placement of STD (submarine tailing disposal) has resulted in a disturbance in the euphotic area that uses oxygen. This is shown by the decrease in the benthos diversity (H' Shannon-Wiener 0.68 to 1.09) and phytoplankton diversity (H' Simpson 0.06 to 0.493) indices, which showed no interference/heavy pollution in the accumulation of tailings in Buyat Bay.

One decade following mine closure, the study demonstrated the dynamics of the marine ecosystem in terms of the macrobenthos diversity. In 2011, the highest diversity was observed at station 5, with 92 genera, while the lowest was observed at station 10, with 9 genera. In 2012, the highest number of genera was at station 4, with 94 genera, while the lowest was at station 7, with 40 genera. In 2013, the highest number of genera was at station 4, with 79 genera, while the lowest was at station 7, with 39 genera. The results showed that the number of genera at the observation stations was closely related to the depth or bathymetric from Buyat Bay. These results are believed to be closely associated with the flow and circulation factor in the waters of Buyat Bay. Therefore, the presence of the types of biota and macrobenthos genera indicates a linkage between the environment and strategies of macrobenthos to survive.

Table 10.1 and Fig. 10.3 show the abundance of the genus/species per taxa of macrobenthos at nine observation stations, which generally indicate an increase in the abundance of the genus/species per taxa.

Figure 10.3 shows the results of observations in 2011 compared to the average abundance. Stations 5, 4, 3, 9, and 1 had a high abundance that was closely related to the number of genera, whereas stations 6, 7, and 8 had an abundance that was not comparable or proportional to the number of genera. Station 10 had a low abundance, which was proportional to the number of genera. Observation of the average abundance in 2012 showed that stations 9, 1, 4, 5, and 3 had a high abundance that was closely related to the number of genera, whereas stations 6, 8, and 10 had abundance values that were not comparable or proportional to the number of genera. Station 7 had a low abundance that was proportional to the number of genera. The total abundance of macrobenthos in Buyat Bay in 2013 varied between the stations; the total abundance was highest at station 4 and lowest at station 10. The six stations with a fairly high abundance were 4, 6, 9, 1, 3, and 5. This indicates that environmental conditions, both chemical and physical factors, as well as the biology of Buyat Bay waters significantly contribute to the life cycle and reproduction of biota macrobenthos. Station 10 is an edge with a sand substrate that is not a suitable habitat for macrobenthos, resulting in a low abundance.

The macrobenthos species diversity index indicated the macrobenthos community characteristics and described the level of diversity of macrobenthos contained in Buyat Bay community. The macrobenthos index diversity values in Buyat Bay waters ranged from 4.40 to 5.35 and included high diversity (>3). The value of the macrobenthos diversity was highest for station 4 and lowest for station 10 (Fig. 10.4).

Figure 10.5 shows the macrobenthos uniformity index in Buyat Bay in terms of a comparison between the value of diversity to the logarithm of the number of species that were present. Based on the criteria, the value of uniformity of macrobenthos in Buyat Bay was high (0.84–0.89).

The macrobenthos dominance index in Buyat Bay is normalized to a dominance of one, and there were a few or many species, which sometimes had a very low value (0.034–0.077), at all sampling stations (Fig. 10.6). Based on the criteria according to Simpson dominance index, none of the macrobenthos species were dominant in Buyat Bay. This indicates that Buyat Bay water is excellent for all macrobenthos types.

The results showed that the number of genera is closely related to the water depth of Buyat Bay; it also appears to be closely associated with the flow and circulation factor in Buyat Bay water. The result of current modeling shows the

Table 10.1 Abundance of the genus/species macrobenthos per taxa in Buyat Bay (individual/M^2)

Stations	*Polychaeta*			Crustacea			Mollusca			Others			Total		
	2011	2012	2013	2011	2012	2013	2011	2012	2013	2011	2012	2013	2011	2012	2013
1	92	1542	1514	49	365	202	30	254	115	9	166	101	180	2327	1932
3	130	1343	1400	106	276	211	208	35	34	16	193	210	460	1847	1855
4	193	1452	1993	65	346	287	44	248	193	22	124	148	324	2170	2621
5	361	1171	1137	122	323	302	224	199	200	44	206	179	751	1899	1818
6	43	738	1887	12	140	172	7	70	93	17	127	260	79	1075	2412
7	45	347	1094	19	95	135	1	37	63	12	37	250	77	516	1542
8	30	584	1171	5	180	181	4	92	91	17	178	154	56	1034	1597
9	109	2533	1720	48	188	222	35	37	48	17	181	271	209	2939	2261
10	12	765	202	1	230	239	0	8	34	6	137	162	19	1140	637

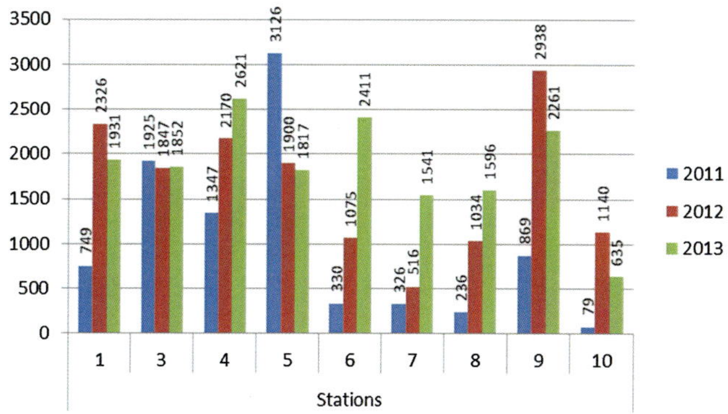

Fig. 10.3 Abundance genus/species macrobenthos per taxa in Buyat Bay

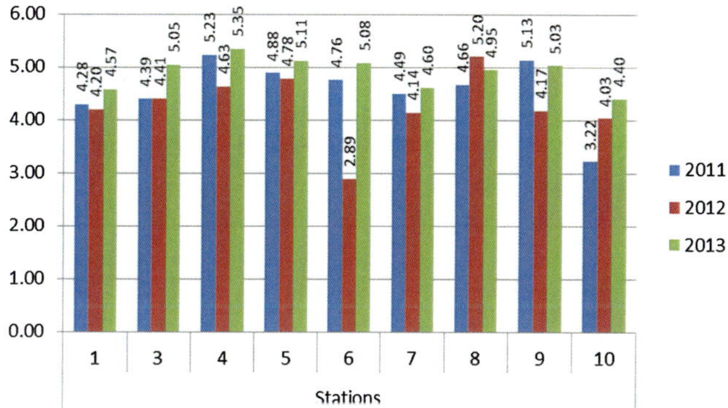

Fig. 10.4 Diversity index for macrobenthos in Buyat Bay

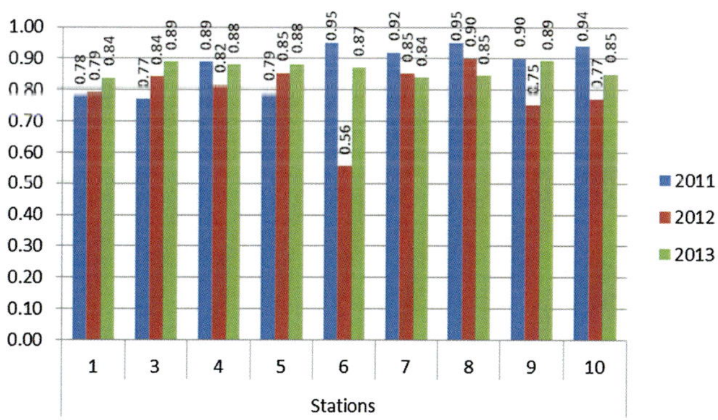

Fig. 10.5 Uniformity index for macrobenthos in Buyat Bay

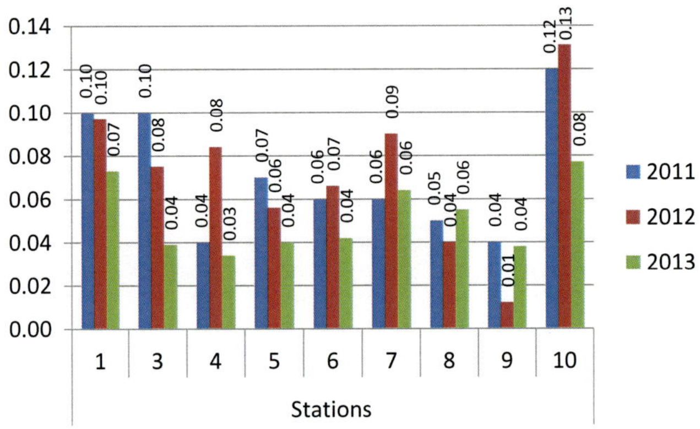

Fig. 10.6 Dominance index of macrobenthos in Buyat Bay

direction of current movement at low tide to the dominant pairs to the north, while at high tide, the current movement reverses to the south. Current velocities occurring in Buyat Bay waters at low tide are ranging from 0.04 m s to 0.08 m/s, while the current velocity occurring during high tide is 0.02 m/s up to 0.06 m/s.

The speed of tidal currents near the coast is relatively low compared to the tidal current rate offshore. This is because the depth of waters near the coast is shallower than the depth of waters offshore. Current velocity is strongly influenced by the depth of the water. The shallow water depth causes the basic friction to become larger.

Based on the description, it is known that the tides of Buyat Bay waters are generated by the difference of tidal amplitudes between the waters of western part of Buyat Bay and the waters to the east. The current velocity at the time of the tide to the ebb is smaller than the current velocity at low tide. This will further assist the organisms that originate from the more outer waters in Buyat Bay carried into Buyat Bay area where the tailings are located.

Therefore, the observed types of biota and genera of macrobenthos indicate a link between the environment and opportunities for macrobenthos biota to maintain their viability. The environmental conditions, both chemical and physical factors, as well as biology of Buyat Bay water significantly contribute to the life cycle and reproduction of macrobenthos biota. There were some genera and species that had not been found before, showing that the bay water offered a suitable environment with available nutrients, especially organic material that directly or indirectly affect the lives macrobenthos. The diversity, uniformity, and dominance showed that Buyat Bay water is in a state of increasing dynamics. The environmental carrying capacity of the water after the disposal of tailings tends to keep the various macrobenthos in equilibrium.

The diversity, uniformity, and dominance showed that Buyat Bay water is dynamic and improving. The environmental carrying capacity of this water tends to provide equilibrium for various macrobenthos. Strategies for the control region (occupying area) are indicated by biota macrobenthos that occupy and live in the sediment. The biota distribution had the same pattern as that of the bottom surface. The benthic community structure can be changed according to the water depth (Clarke et al. 1993), the condition of the substrate water (Johnson and Frid 1995), and the topography that influenced water circulation (Chou et al. 1999).

The observed effects can be caused by the stirring of waves and currents. This circulation provides better nutrients. These conditions impact the increased primary productivity in terms of the dissolved oxygen through photosynthesis and availability of food for marine life. There were some genera/species that were not previously found, especially *Polychaeta*, which increased by nearly 50%. The presence of *Spionidae* and *Capitellidae* indicated the suitability of the environment and availability of food, especially organic material that directly or indirectly (through the presence of animal food for predators) supports *Polychaeta*. For some types of macrobenthos, such as *Polychaeta* and crustaceans, that have a larger size, improvements in the environment supported them.

10.4 Conclusion

Marine macrobenthos communities were affected by tailings. The benthic community structure can be changed by changing the substrate conditions. The macrobenthos community undergoes recolonization in response to tailings several years after mine closure. This illustrates the stages of succession and indicates the link between the environment and opportunities for macrobenthos biota to maintain their viability.

The environmental conditions, both chemical and physical factors, as well as the biology of Buyat Bay waters significantly contribute to the life cycle and reproduction of biota macrobenthos. The presence of newly discovered genera and species was indicative of a suitable environment and nutrient availability, especially organic material that directly or indirectly affected the lives of macrobenthos.

The value of diversity, uniformity, and dominance showed that Buyat Bay water is in a state of improving dynamics. The environmental carrying capacity of the water after the disposal of tailings tends to provide various macrobenthos with equilibrium.

References

Chou WR, Lai SH, Fang LS (1999) Benthic crustacean communities in waters of southwestern Taiwan and their relationships to environmental characteristics. Acta Zoologica Taiwanica 10 (1):25–33

Clarke KR, Warwick RM, Brown BE (1993) An index showing, breakdown of seriation, related to disturbance, in a coral-reef assemblage. Mar Ecol Prog Ser 102:205–219

Edinger, Siregar, Blackwood W (2007) Heavy metal contamination of shallow marine sediments from submarine tailings disposal and artisanal gold mining, Buyat-Ratatotok district, North Sulawesi, Indonesia. 40:404–425

Ellis DV (2008) The role of deep submarine tailing placement in the mitigation of marine pollution for coastal and island Mines. In T. Hover. Nova Science Publisher

Ellis D, Ellis K (1994) Very deep STD. Mar Pollut Bull 28:472–476

Emery WJ, Thomson RE (2001) Data analysis methods in physical. Elsevier

Johnson LL, Frid CLJ (1995) The recovery of benthic communities along the county Durham coast after cessation of colliery spoil dumping. Mar Pollut Bull 30(3):215–220

Jones SG, Ellis DV (1995) Deep water STD at the misma gold and silver mine, Papua New Guinea. Mar Georesour Geotechnol 13:1553–1562

Krebs (1989) Ecological methodology. Harper & Raw, New York

Ministry of Environment Report on Handling research suspected cases of pollution and/or destruction of the environment in Buyat Village Beach and Village East Ratatotok. Ratatotok District of South Minahasa district, North Sulawesi

Odum PE (1993) Fundamental of ecology. Gadjah Mada University Press

Poling G (2002) An introduction to deep sea tailing placement. In: Poling GE (ed) *Underwater tailing placement at Island Copper mine: A success* story. Society for Mining, Metallurgy, and Exploration, Littleton, Colorado, pp 1–16

Prisetiahadi K, Yanagi T (2008) Seasonal variation in the behavior of tailing wastes in Buyat Bay, Indonesia. Mar Pollut Bull 57:170–181

Salomons W (1995) Environmental impact of metals derived from mining activities: process, prediction, prevention. J Geochem Explor 52:5–23

UNCLOS (1982) United Nations Convention on The Law of The Sea. United Nations

Chapter 11
Reinventing Fiscal Policy on a Quintuple Helix Perspective Toward Indonesia-World Maritime Interaction: A Case in Batam Free Trade Zone and Lamongan, Indonesia

Haula Rosdiana, Inayati, and Maria R.U.D. Tambunan

Abstract The aim of this study is to examine current fiscal policy and whether it has been applied effectively and then to examine the possibility of its revitalization to support the Indonesian government vision, with Indonesia as an axis of the world maritime industry. The current situation shows that a large financing gap exists for the provision of maritime infrastructure and that the government has offered fiscal incentives to ease the burden of the maritime industry, even though fiscal incentives are not the best approach in terms of providing for the needs of a business entity. In order to provide comprehensive analysis, the quintuple helix theory was utilized, since it covers five different perspectives in examining a problematic situation: economic, politics, media, education, and environment. This study was conducted using a qualitative approach; the data was gathered through a literature study, in-depth interviews, and focus group discussions. The results show that even though a tax incentive is not the main consideration for enterprises to invest in the marine industry in Batam, for example, in shipyard building and repair, however, it contributes indirectly to the economic activity in Batam. On the other hand, these fiscal incentives also bear drawbacks because several industries construct economic transaction schemes solely to get the benefit from incentives by optimizing the loopholes on tax regulations. Therefore, the government needs to reinvent appropriate tax incentives with comprehensive monitoring. On the other hand, in Lamongan, the current local government is pursuing maximization of current fiscal incentives according to its authority and is offering other incentives such as ease of permitting and promotion of local competitiveness. However, the central government, as the main actor who shall actively participate in supporting its vision, "toward Indonesia as a world centre of the maritime industry," has not showed a real role, unless it is in the stipulation of Lamongan as maritime industrial area. Close coordination between local government and central government must be undertaken to enhance this potential.

H. Rosdiana • Inayati (✉) • M.R.U.D. Tambunan
University of Indonesia, Depok City, West Java 16424, Indonesia
e-mail: inayati.hifni01@gmail.com

B. McLellan (ed.), *Sustainable Future for Human Security*,
https://doi.org/10.1007/978-981-10-5430-3_11

11.1 Introduction

This study is intended to examine current fiscal policy applied in Batam and Lamongan, Indonesia, and whether it has been achieved its aim. Further, it is also intended to examine the possibility of its revitalization to support the government vision, "toward Indonesia as a world centre of the maritime industry" through the development of the marine industry. The current situation shows that a large financing gap for the provision of maritime infrastructure exists. This has been worsened by distortive tax policy, for example, discriminative imposition of value-added tax on supply of services related to the port industry, which is the tip of the iceberg of problems of financing policy and non-comprehensive fiscal policy. All of these problems may add a burden on the marine industry.

Batam is a vital industrial area selected by the government as an influential port and main location for the shipyard industry. To enhance the development of the Indonesian industrial sector and to attract investment to the Batam industrial area, the government provides a special fiscal and non-fiscal treatment for the economic activity in Batam by stipulating Batam as a free trade zone since the 1980s. Indeed, the shipyard industry should get benefit from the provision of these incentives. With regard to this status, the enterprises in Batam receive favorable tax treatment: exemption from the imposition of tax on consumption such as value-added tax, import duty, excise, and prepaid tax (art. 22). However, it is important to examine the current policy applied in Batam to determine whether it has adequate enough support to achieve the government goal and to find other factors that may affect this industry. On the other hand, Lamongan is a new expanded region to support the development of shipyard building and repair by constructing a specific economic zone. In this specific economic zone, a variety of fiscal and non-fiscal incentives are offered and will be examined in this chapter.

11.2 Quintuple Helix Perspective, Fiscal Incentives, and Research Method

The quintuple helix is an interdisciplinary view that grasps the interaction of disciplines within society in order to promote a cooperative system of knowledge. It has been cited that the quintuple helix (Carayanis 2012, 5):

> Is interdisciplinary and transdiciplinary at the same time; the complexity of the five helix structure implies that a full analytical understanding of all helices requires the continuous involvement of the whole disciplinary spectrum, ranging from natural science (because of the natural environment) to social science.

The five components of the quintuple helix consist of economic, political, education, media, and environment perspectives. The economic perspective covers the interaction of entrepreneurship and economic activities. Then, the economic perspective may be affected by the political system in which the government or

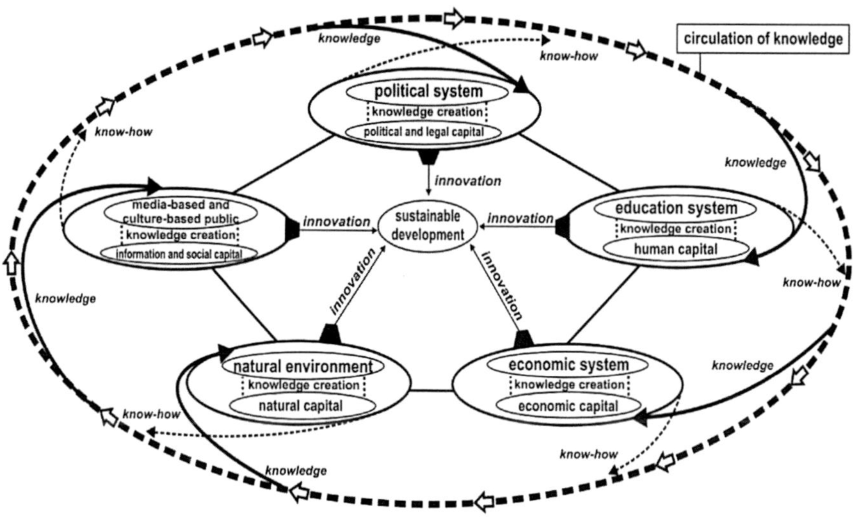

Fig. 11.1 Mode of quintuple helix and its functions (Source: Carayannis, EG, & Campbell, DFJ. Triple Helix, Quadruple Helix and Quintuple Helix and how do knowledge, innovation and the environment relate to each other? A proposed framework for a trans-disciplinary analysis of sustainable development and social ecology. *International Journal of Social Ecology and Sustainable Development* 2010. Hal 61)

state stipulates policies affecting the past, current, and future situation. Media plays a role in publication of information through various kinds of media. The education system refers to human capital and institutions supplying human skills. Finally, the environment shall refer to capital provided by the environment. The quintuple helix model can be drawn as follows (Fig. 11.1).

Fiscal incentives may be defined as special favorable treatments to specific industries by reducing their tax burden with various kinds of scheme. Zee, in Easson (2004) defined tax incentives as:

> In effective terms, a tax incentive would be a special tax provision granted to qualified investment projects that has the effect of lowering the effective tax burden-measured in some way-on those projects, relative to the effective tax burden that would be borne by investors in the absence of the special tax provision.

Tax incentives are intended to reduce total tax burden of investors on specific industries. Then, Easson (2004) proposed that tax incentives can be granted in a number of forms: (a) reduced rates of corporate income tax for particular activities or types of enterprise, (b) tax holidays (i.e., reduction or exemption from tax for a limited duration), (c) investment credits or allowances for investment in capital assets, (d) accelerated depreciation of capital assets, (e) deductions or credits for reinvested profits, (f) reduced rates of withholding tax on remittances to the home country, (g) reduced personal income tax and/or social security contributions for executives and employees, (h) property tax reductions, (i) reduced import taxes and duties, and (j) creation of special zones.

In this research, Batam was selected by the government as a special zone since it is intended to be a center of industrial activity.

This study uses a qualitative research approach. Data and information was collected through literature studies and field research. A series of in-depth interviews with relevant informants and focus group discussions (FGD) were conducted to enrich the information. FGD involved stakeholders consisting of Batam Indonesia Free Zone Authority, Ministry of Finance, Ministry of Transportation, and Researcher/Academicians who were concerned with maritime industry issues. For the Lamongan site, interviews and FGD were also conducted with stakeholders consisting of Lamongan local government, Indonesia Investment Coordinating Board in Lamongan, Lamongan Tax Office, Lamongan Customs and Excise Office, National Planning Agency of East Java, Indonesia Investment Coordinating Board of East Java, and PT Lamongan Integrated Shorebase.

11.3 Batam Free Trade Zone, Shipyard Industry in Batam, and Lamongan Industrial Region

Batam is a free trade zone whose economic activity is organized by a special government body called Batam Indonesia Free Zone Authority (BIFZA). As a specific economic zone, each investment will be granted fiscal and non-fiscal concessions. Fiscal benefits offered to investors as regulated through Government Regulation No.5/2011 consisted of (i) exemption on imposition of customs duty, value-added tax, tax on luxury goods, and excise for supply of goods and services for industrial purposes; (ii) exemption on imposition of customs duty, value-added tax, tax on luxury goods, and excise for supply of goods and services solely to be used in Batam; and (iii) fiscal incentives offered will be available for 70 years from the time of stipulation.

The shipyard industry is a vital industry in Batam because of its considerable contribution to Batam's economic activity. Up to 75% of all Indonesian shipyards are located in Batam (BSOA 2015). Most shipyard entrepreneurs join the shipyard association called the Batam Shipyard and Offshore Association (BSOA). There are 114 out of 250 shipyard entrepreneurs who are members of BSOA. The number of ships produced by members of BSOA continuously increased from 40 units in 2000 to 114 units in 2014.

This type of industry is capital and labor-intensive and is able to absorb a large number of employees. These characteristics also make the shipyard industry a high-risk investment since all the work is contract based, whereby the work is driven by demand of ship from other industries, such as the mining industry to carry its coal product. The main components of ship building are steel as the base material, technology (equipment), and intensive labor. Constructing a new ship will take about 2 years, whereas the price of commodities used like steel and minimum wages tend to change following the economic situation. Unpredictable economic and political situations will severely impact this industry and are worsened by the

Fig. 11.2 Location cluster of shipyard industry in Batam Island (Source: Batam Shipyard and Offshore Association (BSOA) 2015)

difficulties to get loans from banks because of the nature of the shipyard industry. Broadly speaking, any situation may occur within 2 years during the job contract, and economic slowdown can make the shipyards incur losses.

The locational cluster of the shipyard industry in Batam can be seen in the following picture (Fig. 11.2).

The initial concept of establishment of a specific economic zone (SEZ) in Lamongan is to be a new center of the maritime industry other than Batam and Surabaya, the two well-known marine transportation ports. Several considerations to choose Lamongan as a new maritime industrial area were (i) close to Surabaya, an international port in Indonesia, (ii) close to the source of an industrial hub, (iii) located in Java logistic line, (iv) the availability of deep sea and long coast, (v) rich with natural resources, and (vi) close to fish market but far from local settlement. Current integrated maritime industries also exist in Lamongan, consisting of (i) Paciran commodity trade port, (ii) fish and cold storage port, (iii) Lamongan sea tourism, and (iv) Lamongan Integrated Shorebase (LIS). LIS has become the integrated area of shipyard building and repair and industry providing oil and gas logistics. The integration of the industrial area in Lamongan can be seen in the following picture (Fig. 11.3).

1. Fish and cold storage port
2. Lamongan sea tourism

Fig. 11.3 Lamongan Integrated Industrial Area (Source: Lamongan Investment Coordination Board 2015)

3. Paciran commodity trade port
4. Lamongan Integrated Shorebase (LIS)
5. Babam logistic line

Local government in Lamongan has a clear concept of the development of Lamongan maritime industry. They divide Lamongan coast into two industrial regions with different specifications. Brondong is to become the center of shipyard building and repair, whereas Paciran becomes the center of the fishing industry. Local government also pursues development of local competitiveness as a trade commodity. In order to boost the development of the shipyard industry in Lamongan, the government established the National Ship Design and Engineering Center/NSDEC in cooperation with Institute Technology Sepuluh Nopember Surabaya. The industrial region along the north coast of East Java can be seen in the following picture (Fig. 11.4).

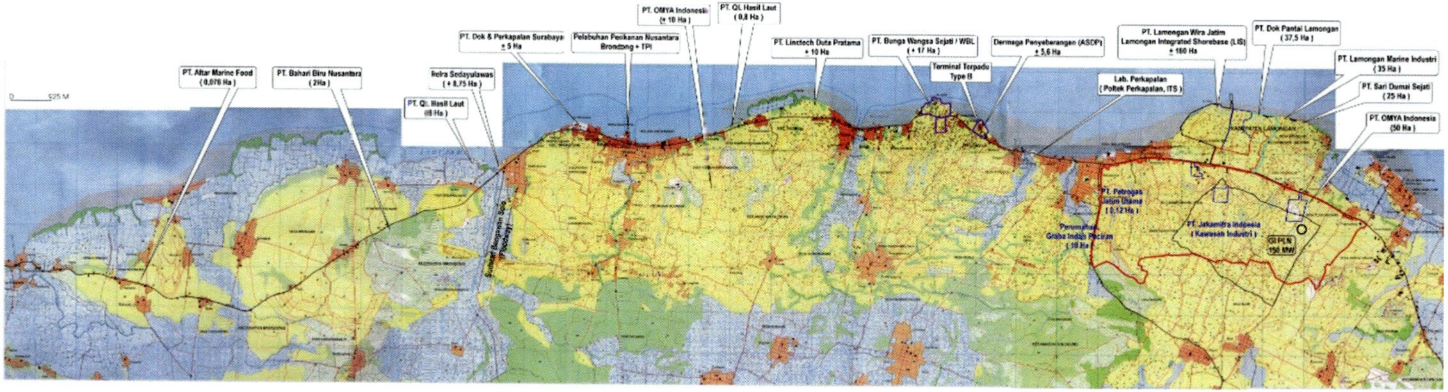

Fig. 11.4 Lamongan maritime industry (Source: Lamongan Investment Coordination Board 2015)

Table 11.1 Investment in shipyard industry, Lamongan

No	Investors	Size	Investment volume(IDR)
1	PT. Dok dan Perkapalan Surabaya	20.5 ha	434,000,000,000
2	Laboratorium Perkapalan, Politeknik Perkapalan Negeri, Institut Teknologi 10 Nopember (ITS), Surabaya	5.5 ha	35,000,000,000
3	PT. Dok Pantai Lamongan	37.5 ha	94,129,800,000
4	PT. Lamongan Marine Industries	35 ha	330,000,000,000
5	PT. Tri Ratna Diesel Indonesia	1.5 ha	30,080,000,000
	Investment volume		923,209,800,000

Source: Lamongan Investment Coordination Board 2015

Until May 2015, the total investment on the shipyard industry in Lamongan was up to IDR. 923,209,800,000. Investment in the shipyard industry is based on information gathered from Lamongan Investment Coordination Board as follows (Table 11.1).

11.4 Reconstructing and Revitalizing Fiscal Policy in Quintuple Helix Perspective

Reemphasizing implemented fiscal incentives, basically the nature of Batam as a free trade zone should ease the investors to maximize their business activities. The latest regulation regarding facilities offered in the free trade zone (FTZ) as stipulated in Government Regulation No. 10/2012 consists of:

(a) Supply of taxable goods into FTZ shall be granted postponement of customs duty, noncollectible VAT, noncollectible prepaid tax (art.22), and/or exemption of excise in accordance to the regulations of bonded stockpiling.
(b) Supply of goods from other FTZ shall be granted noncollectible VAT, exemption from VAT, and/or exemption from excise in accordance to the regulation of bonded stockpiling.

Clear explanations regarding fiscal incentives granted to business entities located in Batam can be seen in the following picture (Fig. 11.5).

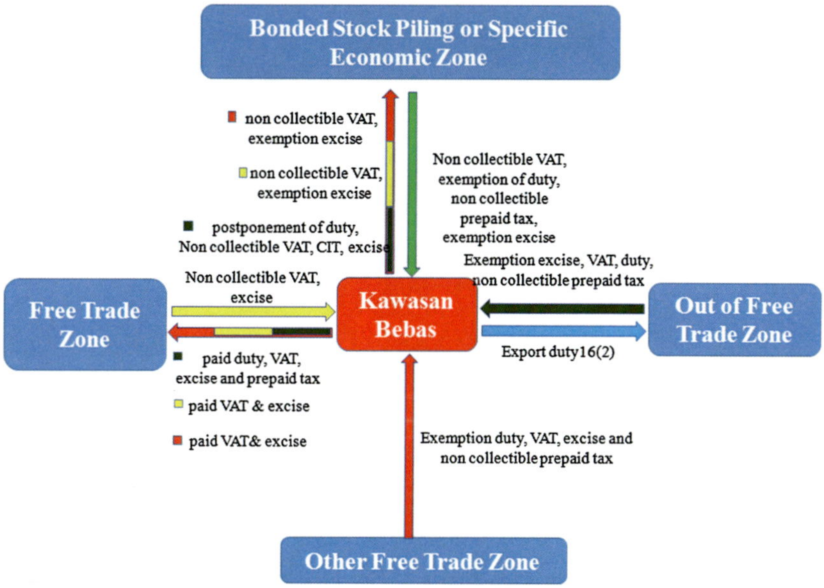

Fig. 11.5 Fiscal incentives scheme on the supply of goods in Batam FTZ

11.4.1 Economic Aspect

Economic activities of the shipyard industry for export orientation happen continuously. Investment in shipbuilding and repair is still attractive, as identified by the number of foreign investors who invest their capital in Batam. While constructing new ships, imported components covers up to 70% of the total. These imported components shall be granted exemption of VAT and customs duty. However, the incentives will not be applicable if the component of new ship is supplied by domestic suppliers, since the shipbuilding is export oriented. Once produced goods are sold out of the FTZ, VAT, import duty, and excise shall be imposed. As stipulated by GR No. 10/2012, taxation on supply of shipyard components in Batam can be drawn as follows (Fig. 11.6).

In order to accelerate the *cabotage* principle which obliges ship sailing in domestic shipping lines to do so under the Indonesian flag, the government exempts imposition of VAT for the supply of new ships domestically and importation of used ship/spare parts in order to increase the number of Indonesian flagged vessels. Tax incentives offered consist of exemption VAT, customs duty, and prepaid tax (art.22). The scheme of tax incentives for the shipyard industry out of the FTZ can be seen in the following picture (Fig. 11.7).

In several business entities' views, the application process to get tax incentives such as VAT and customs duty exemption frequently bears cumbersome administrative requirements. Time spent in fulfillment and verification of documents for incentive application sometimes slows down production processes, since the supply

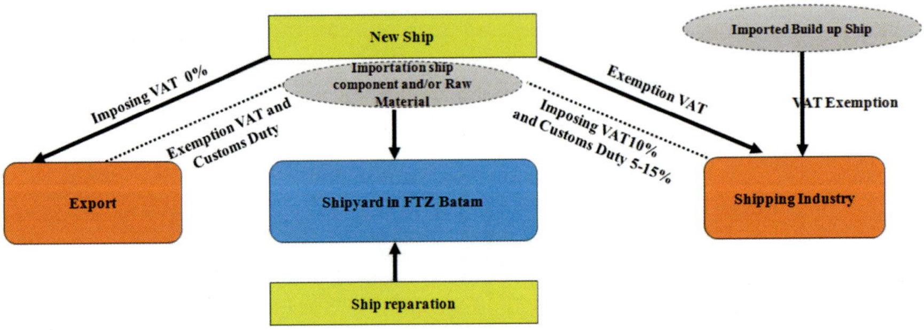

Fig. 11.6 The scheme of tax incentives for shipyard in Batam (Source: Rearrange form GR No. 10/2012)

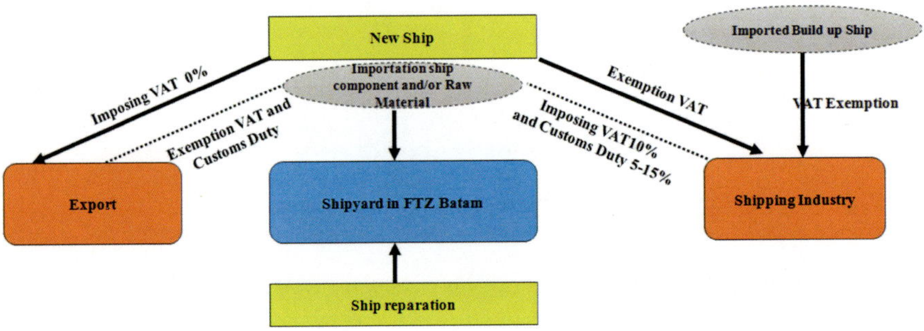

Fig. 11.7 The scheme of tax incentives for shipyard out of free trade zone (Source: Rearranged from GR No. 10/2012)

of goods/equipment to be utilized for industry activities must be released by a competent authority in the relevant institution after a series of audits. A few investors even argue that the tax incentives are not really beneficial for them. For investors, bureaucracy and administrative processes and rule of law also affect them in making decisions. Incentives are not a priority to make their decisions. Besides matters related to the government, the shipyard industry can be improved faster if the local content can supply material or equipment as standardized. The dependence on imported equipment adds expensive transportation costs or logistic supply chains making overall costs more expensive compared to ships produced by other countries such as China. Small and medium enterprises (SMEs) have not played a significant role in the provision of the local content to date.

On a separate note, monitoring the application of fiscal incentives for investment is a challenging task, since these incentives can potentially be utilized in contradiction with the initial government objective. The facts show that fiscal incentives are utilized by business entities to gain benefit by constructing various schemes. This means that it may also be used as an opportunity to get economic rent through

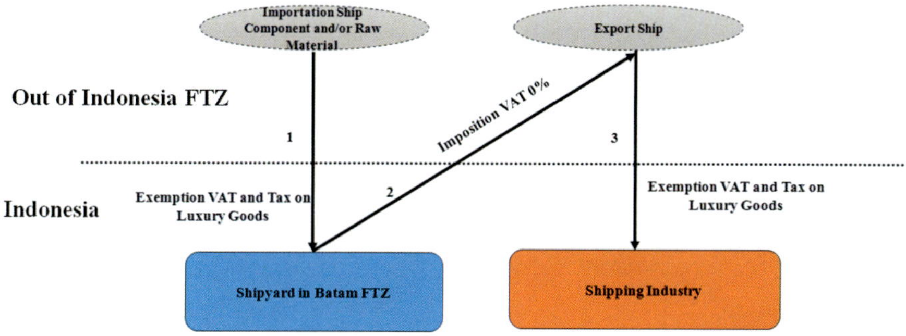

Fig. 11.8 The scheme of tax avoidance performed by shipyard industry (Source: Field Study Finding 2015)

tax avoidance with kinds of transfer pricing schemes as drawn in the next picture (Fig. 11.8).

11.4.2 Political Aspect

Even though the government offers fiscal incentives, the political will of the government in supporting business activities is also important. In addition, fiscal incentives may be considered as complementary to the macroeconomic situation. Besides fiscal instrument, non-fiscal instruments such as the single-window permitting can also make it easier for them to accomplish administrative requirements. From the business entity view, the government has not formulated a clear concept of the maritime industry even though mass media have widely published the government vision "toward Indonesia as a centre of the world maritime industry." By running business as a usual activity, the local government shall be given discretion to administer and organize local economic activity. Authority issues between local and central government will contribute to slowing down the achievement of the Indonesian maritime industry competitiveness.

In addition, lack of coordination among institutions causes overlapping regulations, which frequently raises confusion among business entities trying to make business decisions. The sudden change of regulations without an adequate socialization or public hearing and the dynamics of the political situation will lead to distortion, even though the government offers incentives. Fiscal incentives with easy administrative procedures will be fully utilized if the government is able to ensure a conducive and friendly investment climate. One example of ambiguous regulations is the regulation of urban plan organization which stipulates part of shipbuilding locations become conservation forest after construction activity if they are officially held in a conservation location. The release of this regulation forced entrepreneurs to stop their activities and stopped banks from giving loans.

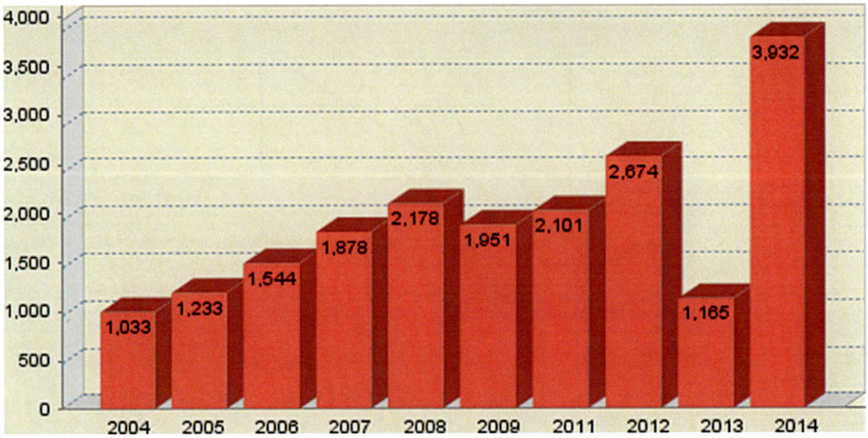

Fig. 11.9 Tax revenue in Batam (in billion rupiahs) (Source: Indonesia Batam Free Trade Zone Authority 2014)

11.4.3 Environmental Aspect

For this context, the environmental aspect may be associated with geographical location. Batam has a strategic position since it lies in an international shipping line. This strategic location is also one of the considerations to establish it as a free trade zone. Since its stipulation as an FTZ, numbers of inhabitants in Batam dramatically increased from 100,000 people to 1,035,028 in 2014. The growth of population is in line with the growth of employment. Based on information gathered from BSOA, up to 70% of employees in Batam work for the shipyard industry. Therefore, it might be said that fiscal incentives indirectly affect job creation in Batam. Further, these large numbers of employees contribute to the government tax revenue through personal income tax. At least 10% of the total tax in Batam comes from the shipyard industry. The following is the information of tax revenue from employment and business activity (Fig. 11.9).

11.4.4 Education Aspect

In most of the country, granting fiscal incentives is intended to adopt skills through transfer of technology. In Batam, since most of investors come from abroad, transfer of technology should be essential, as stipulated in Indonesia Investment Law. However, the facts show that transfer of technology has not been commonly applied. There is no significant impact on technology development among the local people in Batam. In addition, most of the local people work as laborers with low salaries. In one case, construction processes for a new technologically sophisticated ship cannot be observed by employees other than expatriates.

11.4.5 Media Aspect

Media plays a role as disseminator of information from business entities to the government and vice versa. Batam as an FTZ which has been a center of industrial activities commonly becomes a point of interest of the media. In Batam, the media has played their role to promote government policy specifically in realization of the friendly investment climate. Also, it takes a part as a connector of business entities into the government to raise problems of investment.

11.5 Optimizing Fiscal Policy on the Development of Maritime Industry in Lamongan

As a specific economic zone, the government offered various kinds of fiscal incentives and non-fiscal incentives in order to realize a friendly investment climate in Lamongan. Fiscal incentives are offered by both the central government and local government. Fiscal incentives offered by the central government are administered based on Law No. 39 year 2009 regarding specific economic zones. Fiscal incentives consist of (i) income tax incentives with a variety of allowances based on the characteristics of the economic zone and type of business, (ii) tax on land and building incentives, (iii) customs duty/tariff and postponement of import duty, (iv) exemption on excise, (v) uncollectible value-added tax (VAT), and tax on luxury goods. Then, the supply of taxable goods and service into and from specific economic zones for eligible entrepreneurs may be offered incentives, for example, uncollectible VAT.

11.5.1 Economic Aspect

As a specific economic zone, various kinds of incentives have been offered to the industry in Lamongan. However, imposition on spare parts remains a problem, since importation of these components must incur VAT. Most shipyards have to import ship components up to 70% of the total needed. This treatment may sound unfair since shipping industries which import second-hand ships will be exempted from the imposition of VAT. Exemption of VAT on the importation of ship components means some goods used which are subject to VAT cannot be

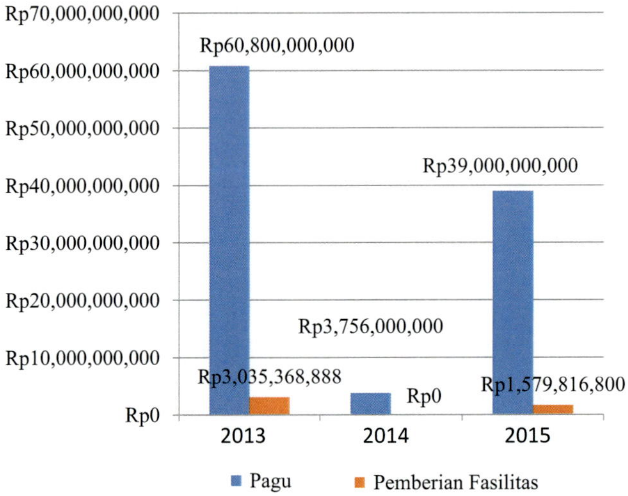

Fig. 11.10 Utilization of duty incentives on shipping industry (Source: Ministry of Finance 2015)

considered in the VAT input-output mechanism. Therefore VAT paid will be shifted into the component price.

In the context of customs facility, this incentive has not been fully utilized by business entities because the detailed regulation regarding the utilization of this incentive was released in a very short time. Then, the procedure to apply incentives is quite complicated. Since 2013–2015, the utilization of duty incentives as recorded by the Ministry of Finance is as follows in Fig. 11.10.

Besides fiscal incentives offered by the central government, the local government also offers fiscal incentives as stipulated in Law No. 39 year 2009 in the form of exemptions, reduction of local tax, and user charges. Specifically, the local government in Lamongan offered local tax incentives and user charges as regulated through local government regulation by granting reduced rates of local tax and some kinds of permit free of charge. Non-fiscal incentives offered consist of the ease of permit license issuance for the acquisition of land, immigration permits, and other important permits based on characteristics of the economic zone and type of industry. Lamongan has established a single-window permitting system to reduce the numbers of permits for setting up new businesses.

11.5.2 Political Aspect

Political will of the government to enhance maritime industry competitiveness by constructing a maritime industry blueprint is an important aspect since it can be a guideline for related entities to optimize their role. From the investor viewpoint, the current local government is cooperative enough and supports business activities in

Lamongan. Local government also provides fiscal incentives, for example, ease of administration along with abundant resources and competitive salaries for labor become important factors for business development. Local government has formulated official guidelines for business license permitting. This guideline gives certainty to business regarding the accomplishment of administrative and permit licensing. In addition, this guideline also enforces that the local government works professionally. This friendly investment climate has made a positive impact on investor's business activities.

Unfortunately, even though business activities in Lamongan show considerable improvement, this result may be due to the local government contribution through their political will to reach higher economic growth and income per capita in Lamongan. Local government also tries to list problems and propose alternative solutions for investment into areas which potentially slow down economic activity. Most of these findings are related to infrastructure availability, for example, permanent roads. Transportation carrying out logistical activities bringing commodities into Lamongan has to pass through unusual roads/lines for passage because of the poor quality of roads. Trucks have to seek alternative roads, which makes transportation and logistics costs become more expensive than normal.

Local government has raised these problems and seeks for solutions from the central government; however, the central government has not provided objective solutions. Besides infrastructure, land clearing also remains a problem, since the cost of land clearing is high, this will be a burden to new investors. This problem shall be intervened with by the central government in order to reduce the cost burden.

The central government has not set a clear concept of the maritime industry, and its target is expected to be reached by offering various kinds of incentives. The central government seems to have simply established the economic zone in Lamongan. Less coordination between the local and central government regarding the road map of this economic zone will probably become a challenge in the future since the local government has limited authority to formulate policy, whereas every policy must meet with the central government's vision.

11.5.3 Education Aspect

The education aspect also plays a significant role in sustainable development since educational institutions are the ones able to provide high-quality workers. The shipyard industry is a capital and labor-intensive sector which needs experts in constructing high technology ships. However, the availability of standardized technology for the shipyard industry remains a problem. Most shipyard industries rely on imported machinery. Even some types of raw material have to be imported because the quality of Indonesian product, for instance, steel, cannot meet the standardized requirement. Besides that, shipyard business entities also have to invite professional expatriates to handle specific high technology jobs, for example, jobs related to welding and construction. Currently, most of the local employees work in low-skilled jobs; however, business entities are pursuing development of the local people's skill by providing internal and external training.

11.5.4 Media Aspect

Media play the role in dissemination of the situation to all stakeholders. This may be viewed as policy dissemination infrastructure or as a tool of information sharing into society. Media also play the role in distributing information or promotion of particular areas to potential investors. In Lamongan, the media take a role as an information sharing tool that can promote government policy specifically in realization of a friendly investment climate. Also, it takes the part of a connector of business entities into the government to raise problems of investment.

11.6 Conclusion

Fiscal policy will be beneficial to develop the investment climate if the administrative matters related to tax incentives can be achieved in a simple way. This can also reduce the real tax burden. The vital role of the government in formulating business-friendly regulation, the dynamic of global economics, supply of equipment and material needed, bureaucracy, and coordination between related institutions and between central and local government play a part in acceleration of maritime industry development. Importantly, the government also has to monitor the utilization of tax incentives, so that the business entities will not optimize the potential loopholes in tax regulation to avoid potential tax through aggressive tax planning such as in Batam. On the other hand, for the Lamongan case, giving more flexible authority to certain extents to the local government may make it easier for the stakeholders to complete their requirements related to government approvals.

References

Badan Koordinasi Penanaman Modal. (2010). Pra Studi Kelayakan Industri Perkapalan: Kajian dan Pemetaan Sektor Unggulan dan Industri Pengembangannya. Jakarta

Carayannis. (2012). The Quintuple Helix Innovation Model: Global Warming as a Challenge and Driver for Innovation. J Innov Enterp

Easson A. (2004). Tax incentives for foreign direct investment. Kluwer Tax International, New York

Government Regulation No. 100 year 2012 regarding the Amendment of Government Regulation No. 2 year 2011a regarding Specific Economic Zone

Government Regulation No. 5 *year* 2011b *regarding the Amendment of Regulation No. 46 year 2007 regarding Free Trade Area and Batam Free Trade Port*

Governor Regulation of East Java No. 6 year 2012 The Management of Coastal Area and Small Islands year 2012–2032

Governor Regulation of East Java No. 6 year 2012 The Management of Coastal Area and Small Islands year 2012–2032

In-depth interview with Batam Authority

In-depth interview with Indonesia Investing Coordinator Board of Batam

In-depth interview with Directorate General of Customs and Excise Batam

In-depth interview with Directorate General of Taxes Batam

In-depth interview with Batam Shipyard and Offshore Association

In-depth interview with Ministry of Transportation Batam

IPERINDO. (2014). Shipbuilding Industry in Indonesia. Jakarta

Lamongan Local Government Regulation No. 2 year 2010 regarding Port Management in Lamongan

Ministry of Transportation Regulation No. KM 61 year 2008 regarding Master Plan of Port

Presidential Stipulation No. 22 year 1986 regarding Bonded Zone

Presidential Stipulation No. 21 year 1987 regarding The Procedure of Goods Supply and Exit from and into Batam Bonded Zone

UNCTAD (2012) Review of maritime transport 2012. New York/ Geneva, United Nations. www.unctad.org/en/Pages/Statistics.aspx

World Bank IFC (2013) Doing Business in Indonesia 2012

Chapter 12
Seaweed Utilization for Phytoremediation of *Litopenaeus vannamei* Shrimp Farming Waste in Recirculation Systems (Environmentally Friendly Design of Sustainable Shrimp Culture)

Sinung Rahardjo, A. Harsono Soepardjo, D. Djokosetiyanto, and Abimanyu T. Alamsyah

Abstract This research aims to create a model of shrimp culture applications using seaweed as phytoremediation in recirculation systems. The research is expected to solve issues of environmental damage to the exploitation of mangrove forests for shrimp culture area and culture waste disposal into open waters. The data showed that damage to the environment and the spread of various diseases have led to crop failure and reduced production of shrimp culture in several countries including Indonesia. To solve these problems, we need a shrimp culture technology incorporating an ecological approach or in other words eco-friendly shrimp culture *(environmentally friendly)* with waste utilization as optimal as possible. In addition, these technologies should be inexpensive and easily applied by society and economically profitable. Seaweed as well as biological filters can also increase the value of aquaculture production. The increasing value of the benefits of waste will have a positive impact on the use of resources more efficiently. The research data showed that *Gracilaria* sp. has the ability to utilize organic waste from shrimp farming *Litopenaeus vannamei* which was higher than *Eucheuma* sp. and *Caulerpa* sp. Field-scale model test results show that the use of seaweed is also able to increase the growth of shrimp culture and water quality. So, this technology is able to reduce the level of water pollution and maintain the sustainability of the production and aquaculture activity.

S. Rahardjo (✉) • A.T. Alamsyah
Environmental Science Program, University of Indonesia, Jl. Salemba Raya no. 4, Jakarta - 10430, Indonesia
e-mail: snngrahardjo@gmail.com

A.H. Soepardjo
Physics Department, Faculty of Mathematics and Natural Sciences, University of Indonesia, Depok 16424, Indonesia

D. Djokosetiyanto
Faculty of Marine and Fisheries, Agriculture Institute of Bogor, Jl. Dramaga, Bogor, Indonesia

B. McLellan (ed.), *Sustainable Future for Human Security*,
https://doi.org/10.1007/978-981-10-5430-3_12

12.1 Introduction

Mangrove forest area has been decreased significantly from about 4 million ha in 1982 to only 2.5 million ha in 1993. Furthermore, the degradation rate of mangrove forests in Indonesia reached 200,000 ha/year. This has occurred due to the increase of mangrove forest conversion to other purposes such as clearing ponds, industrial and residential area development in coastal areas and clearing of mangrove forests for firewood, charcoal and building materials. The conversion of mangroves into aquaculture ponds occurred widely in North Sumatra, Lampung, West Java, Central Java, East Java and South Sulawesi. Meanwhile, the conversion of mangrove land into industrial zones and residential areas has been common in densely populated areas such as Jakarta, Tangerang and Bekasi, West Java, Central Java, East Java and East Lampung. Direct costs for restoration due to the exploitation of shrimp pond ecosystem in ASEAN ranged from 5.5 to 7.7 billion US $/year which was larger than export earnings (Dahuri et al. 2001).

Shrimp culture activities in the coastal areas can often be one of the causes of destruction of mangrove forests and declining water quality caused by the waste of aquaculture disposal operations. Shrimp culture in Indonesia has contributed quite significantly to the total fisheries production. Since 2002–2007, the contribution of total fisheries production was about 31% per year and 65% for the value of production (Directorate General of Aquaculture. Masterplan Brackish Aquaculture Development in Indonesia 2003).

However, the high volume and valuc of shrimp production will have adverse impacts both directly and indirectly for the environment. This has been indicated by the decline of the shrimp production rate since 1990. The waste from intensive shrimp culture is naturally dependent on the level of assimilation of waste. Roughly 7.4 to 21.6 times the land area used in shrimp culture is required to assimilate waste from shrimp culture. The impacts of aquaculture activities include decreasing of the carrying capacity, water quality and the increasing of disease outbreaks (FAO. Reducing environmental impacts of coastal aquaculture 1991). The application of a technology and management in shrimp culture is intended to increase production by stabilizing the environment changes.

The main cause of environmental degradation in intensive culture systems is the high level of input which is required to support high production outputs. During the culture period, a large proportion of commercial feeds containing high protein content are not assimilated by the shrimps and settle to the bottom of the pond as pollutant material. Approximately 10% of the feed is dissolved and 15–50% may remain uneaten (Lin et al. 1993). The remaining 75% is ingested, but 50% is excreted as metabolic waste, producing large amounts of gaseous, dissolved and particulate waste (Troell et al. 1999). Subsequently, the pond effluent contains elevated concentrations of dissolved nutrients (primarily ammonia), plankton and other suspended solids (Alex and Theresa 2000). The dissolved nutrients and organic material in shrimp ponds stimulate rapid growth of bacteria, phytoplankton and zooplankton (Troell et al. 1999). These accumulated materials may enhance eutrophication, hyper-nitrification and organic enrichment (Lin et al. 1993); (Pillay 2004) which generate unsuitable water quality for black tiger shrimp, leading to disease outbreaks or even mass mortality.

Litopenaeus vannamei is native to the western coast of Latin America. In Indonesia, it is a substitute for tiger shrimp (*Penaeus monodon* Fab.) which were subjected to various constraints of disease and crop failure. The demand of shrimp product has increased over the years. In 2013, the government expects the national shrimp production to have reached 608,000 tons, consisting of 148,500 tons of black tiger shrimp and *L. vannamei* 459,500 tons. It rose by 14% to 699,000 tons, comprising 188,000 tons of black tiger shrimp and 511,000 tons of *L. vannamei* in 2014. Meanwhile Shrimp Club Indonesia (SCI) targeted the production in 2014 amounting to 150,000 tons or 20% increase from the previous year. The increase in demand has caused the public to be encouraged to increase the production by trying various technologies. For example, a shrimp farmer from South Sulawesi successfully increased the productivity of shrimp culture ponds from 9 tons per 1000 m^2 to 15 tons per 1000 m^2. Proper waste management is required to minimize the negative impact of shrimp farming to the surrounding environment.

12.2 Material and Methods

12.2.1 Experimental System

This research was performed in Karangantu, City of Serang, Banten Province. This is applied research with three continuing stages, namely:

1. Step one: The main purpose of this research is to analyse the effectiveness of several kind of seaweeds including *Gracilaria* sp., *Eucheuma* sp. and *Caulerpa* sp. as phytoremediation candidates for utilization of *L. vannamei* farming waste. Three fibreglass tanks (2 m^3 in volume) were used as rearing facilities for shrimp. In addition, nine square tanks with volume 2 m^3 were used as rearing tanks for seaweed. The shrimp density was 300 shrimp/m^3 with size 6 g/shrimp, and density of seaweed was 10 g/L. The research lasted 45 days. Standard shrimp feed with protein content of 50% were given at 4% of biomass five times per day (Fig. 12.1).
2. Step two: The purpose of this research was to analyse the effectiveness of a shrimp culture technology model with a recirculation system combined with seaweed phytoremediation at field scale in terms of technical, ecological, social and economical aspects. This research was conducted using six wooden tubs covered by plastic of size 2 × 7 × 1.5 m for shrimp and seaweed maintenance. The shrimp density was 300 shrimp/m^3 with average weight 5 g/shrimp. An axial water pump was used to help the recirculation system pump the waste from culturing to the maintenance seaweed tub (Fig. 12.2).

Fig. 12.1 Experimental design of Step I, (**a**) *Caulerpa* sp., (**b**) *Gracilaria* sp., (**c**) *Eucheuma* sp.

3. Step three: the aim of the research was to determine people's perception on the technology being tested. Respondents included fish farmers of P2MKP group (Marine and Fisheries Community Training Center) in Minapolitan, Domas Village, Province of Serang, Banten. Purposive sampling method was used to identify and inventory respondents' answers on the questionnaire. Interviews were done to delve deeper into the basis of respondents' answers. Data based on the answers were demonstrated in percentages.

12.2.2 Data Collection

Primary data which was collected includes (1) water quality data (nitrate, nitrite, ammonia, DO, pH and temperature), (2) growth data of shrimp and seaweed, (3) data on survival rate of shrimp, (4) the production data of shrimp and seaweed, and (5) the perception and response of society. Nitrate, nitrite and ammonia were measured every week, while DO, pH and water temperature were measured every day. Seaweed and shrimp growth was measured every week, while the survival rate and productivity were calculated after harvesting.

Fig. 12.2 Experimental design of Step II

12.2.3 Data Analysis

All data collected were analysed using descriptive statistics to assess the effectiveness of shrimp farming models with environmentally friendly recirculation sustainable system. Calculation of biofiltration efficiency was made using the following formula:

$$\text{Efficiency of Biofiltration } (\%) = \frac{(A - B)}{A} \times 100$$

Where:

A = concentration (NH_3, NO_2 and NO_3) at the inlet of seaweed tank.
B = concentration (NH_3, NO_2 and NO_3) at the outlet of seaweed tank.

Average daily growth (ADG) calculated using the following formula:

$$\text{ADG (g/day)} = \frac{(W_t - W_0)}{t}$$

W_t and W_o were, respectively, the weight of seaweed or shrimp at the end and the beginning of the study, and t is the length of culture (days).

Economical and social analysis was used to explain the benefits and perceptions of people regarding the technology being tested to be applied in the field. Assumptions were made to explain the use of the applied models both socially and economically. Evaluation on employment, additionaal income of seaweed-derived businesses, and resulting environmental control value was used in this analysis.

Data measurement from monitoring of waste materials from shrimp culture in intensive ponds was used as comparison.

12.3 Results

12.3.1 Potential Seaweed Candidates

The research results indicated that *Gracilaria* sp. has a greater filtration ability than that of *Caulerpa* sp. and *Eucheuma* sp. in absorbing nitrates, nitrites and ammonia. Based on Fig. 12.1, the level of *Gracilaria* sp. absorption against ammonia, nitrite, nitrate and total organic matter 60–130% was higher than that of *Caulerpa* sp. and *Eucheuma* sp.

12.3.1.1 Seaweed Biofiltration of Ammonia

Biofiltration level of *Gracilaria* sp. for ammonia was the highest in comparison with other types of seaweed, showing a range of 16.7–50.0% followed by *Caulerpa* sp. with a range of 15.3–37.0% and *Eucheuma* sp. with a range of 12.2–35.0%. Average biofiltration rate of *Gracilaria* sp., *Caulerpa* sp. and *Eucheuma* sp. was 36.3%, 24.8% and 20.1%, respectively. A graph of biofiltration level is shown in Fig. 12.3.

ANOVA test results showed that there was a distinct difference in the level of biofiltration of three different types of seaweed as opposed to its ammoniac level. This can be seen from F value (9.406) > F bTable (5.14) confidence level 0.05 (95%). Meanwhile, the Tukey test results showed that there was a difference in

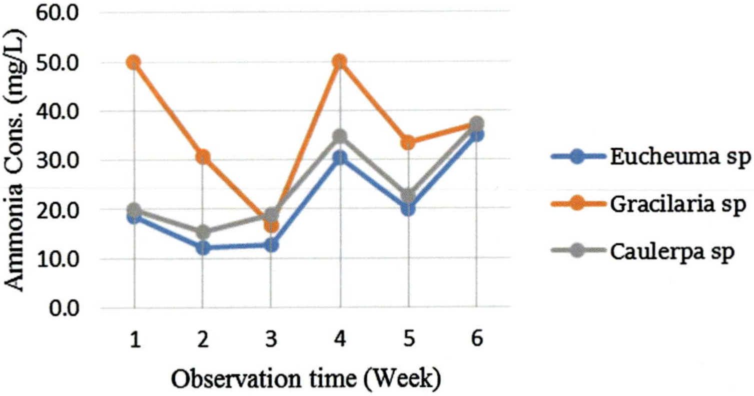

Fig. 12.3 Graphic of biofiltration level of *Eucheuma* sp., *Gracilaria* sp. and *Caulerpa* sp. against ammonia

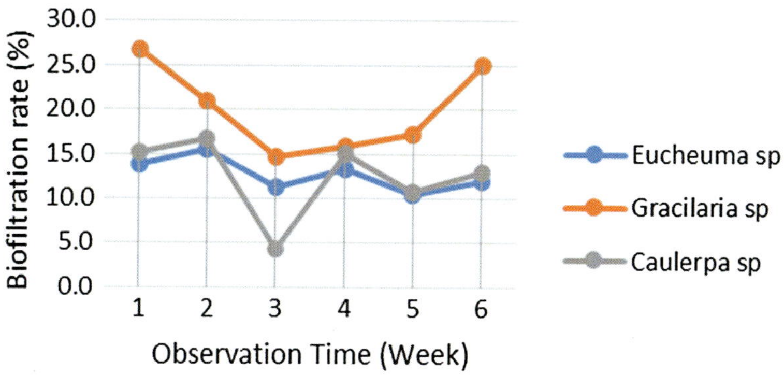

Fig. 12.4 Biofiltration level graph of *Eucheuma* sp., *Gracilaria* sp. and *Caulerpa* sp. on nitrite

biofiltration level of *Eucheuma* sp. and *Gracilaria* sp. against ammoniac level. However, the results did not show any difference with *Caulerpa* sp.

12.3.1.2 Seaweed Biofiltration of Nitrite

Biofiltration level of *Gracilaria* sp. on nitrite was the highest in comparison with other types of seaweed, showcasing a range of 11.1–26.7% followed by *Caulerpa* sp. with a range of 4.3–21% and *Eucheuma* sp. with a range of 1.3–25%. A graph of biofiltration level is shown by Fig. 12.4.

ANOVA test results revealed that there was a distinct difference in the level of biofiltration of three different types of seaweed on the absorption of nitrite. This can be seen from F value (18.9) > F bTable (5.14) confidence level 0.05 (95%). Meanwhile, the Tukey test results showed that there was a difference in biofiltration level of *Eucheuma* sp. and *Gracilaria* sp. against nitrite level. However, the results did not show any difference with *Caulerpa* sp.

12.3.1.3 Seaweed Biofiltration of Nitrate

Average level of biofiltration of *Gracilaria* sp. is 17.7% which was higher than *Eucheuma* sp. and *Caulerpa* sp. with only 12.3% and 8.9%, respectively. Average biofiltration level of each seaweed is shown in Fig. 12.5.

ANOVA test results using SPSS16 revealed that there's a distinct difference in the level of biofiltration of three different types of seaweed as opposed to its ammoniac level. This could be seen from F value (25.5) > F bTable (5.14) confidence level 0.05 (95%). Meanwhile, the Tukey test results showed that there was difference in biofiltration level of *Eucheuma* sp. and *Gracilaria* sp. of nitrate against ammoniac level. However, the results did not show any difference with *Caulerpa* sp.

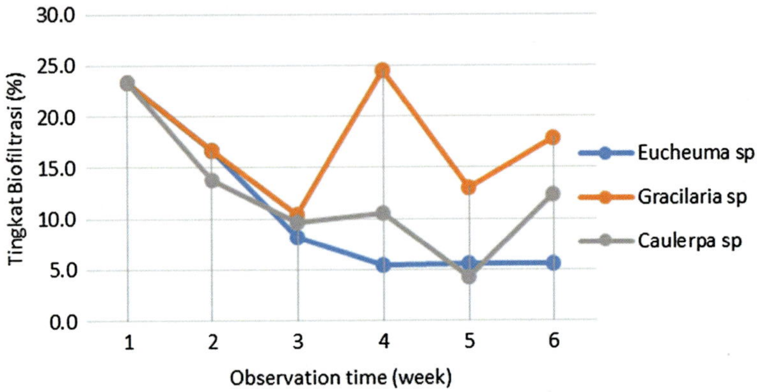

Fig. 12.5 Biofiltration level graph of *Eucheuma* sp., *Gracilaria* sp. and *Caulerpa* sp. on nitrate

12.3.1.4 Seaweed Biofiltration of TOM

Biofiltration level of *Gracilaria* sp. on TOM was the highest in comparison with other types of seaweed, showcasing a range of 12.0–16.8% followed by *Caulerpa* sp. with a range of 3.2–15.0% and *Eucheuma* sp. with a range of 4.5–10.6%. Average biofiltration rate of *Gracilaria* sp., *Caulerpa* sp. and *Eucheuma* sp. was 14.04%, 9.91% and 8.06%, respectively. The graph of biofiltration level is shown in Fig. 12.6.

ANOVA test results revealed that there's a distinct difference in the level of biofiltration of three different types of seaweed as opposed to its ammonia level. This could be seen from F value (9.406) > F bTable (5.14) confidence level 0.05 (95%). Meanwhile, the Tukey test results showed that there was difference in biofiltration level of *Eucheuma* sp. and *Gracilaria* sp. against ammoniac level. However, the results did not show any difference with *Caulerpa* sp.

12.3.2 Application of Field-Scale Model

12.3.2.1 Growth

The results showed that the growth rate of *L. vannamei* was 0.1–0.6 g/day with an average weight range between 15.9 and 16.78 g after 45 days of rearing. The growth rate achieved more than the average normal standard growth of *L. vannamei* that is 0.2 g/day. Growth of each tank is presented in Table 12.1, and the chart of shrimp growth is presented in Fig. 12.7.

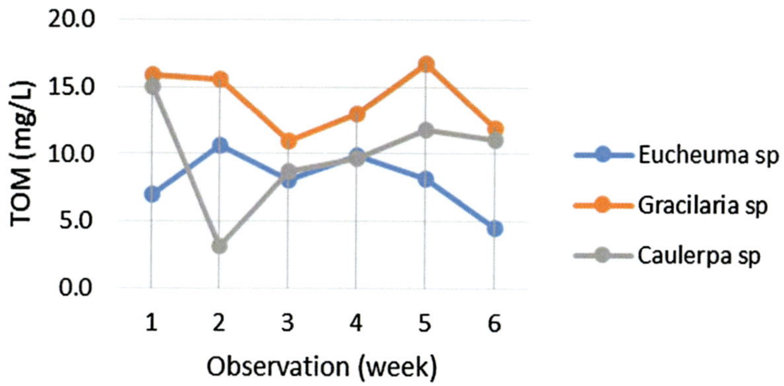

Fig. 12.6 Biofiltration rate of *Eucheuma* sp., *Gracilaria* sp. and *Caulerpa* sp. on TOM

Table 12.1 Measurement growth data of *L. vannamei* shrimp during research

Sampling	Tank A (g)	Tank B (g)	Tank C (g)	ABW (g)	ADG (gDay^{-1})	Biomass (kg)
1	3	3	3	3	0.1	14
2	5.77	5.77	5.77	5.8	0.1	26
3	6.74	7.385	7.359	7.2	0.2	32
4	11.3	11.14	11.76	11.4	0.6	50
5	16.46	14.86	15.53	15.6	0.6	62
6	16.49	16.78	15.91	16.4	0.1	66

Note: *ABW* average body weight, *ADG* average daily growth

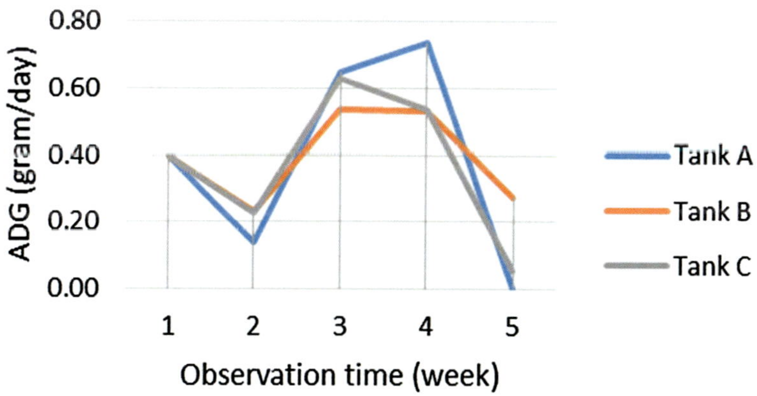

Fig. 12.7 Graph of ADG shrimp during the research

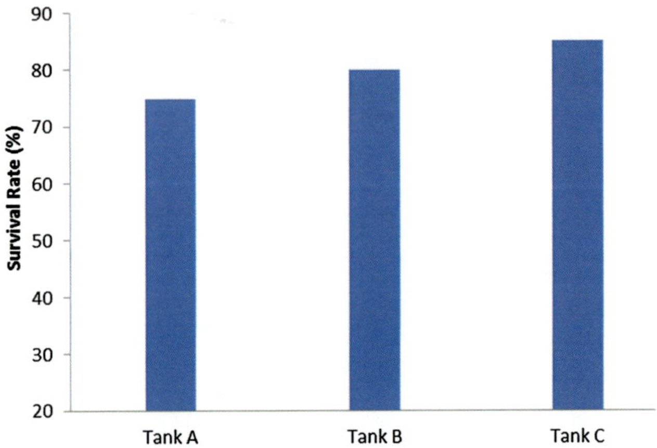

Fig. 12.8 Survival rate of shrimp during the research

12.3.2.2 Survival Rate

Survival rate of shrimp ranged between 75 and 85%. A chart of survival rate during the research was presented in Fig. 12.8.

12.3.3 Social Perspective

Based on questionnaires which were distributed to shrimp farmers, 61% of respondents absolutely agree that the model improved societal income, reduced pollution, was easy to apply, would be able to stop the felling of mangroves, would be able to utilize marginal lands and would be able to maintain the sustainability of production, whereas 20% agree, 13% less agree and 6% did not agree (Table 12.2).

12.4 Discussion

The incorporation of seaweed species in the recirculation system of shrimp farming is a promising environmentally friendly shrimp culture technology which is required in order to run sustainable shrimp farming. Problems such as shrimp disease have occurred in the past most likely due to less concern from farmers in controlling the waste. Phytoremediation using *Gracilaria* sp. is proven to be a solution to increase the absorption of organic waste resulting from shrimp culture activity, especially ammonia and nitrite (see Fig. 12.1). Those two compounds (of ammonia and nitrite) are considered to be harmful compounds that affect growth and survival of shrimp even in the low concentrations. Gills of shrimp more quickly

Table 12.2 Perception of respondent about the application of model technology

Questions	Absolutely agree (%)	Agree (%)	Less agree (%)	Don't agree (%)
Technology can improve people's income	0.9	0.1	0.0	0.0
Technology can reduce the pollution levels	0.6	0.2	0.1	0.1
Technology is easy to be implemented/applied in the shrimp farmer community	0.6	0.2	0.2	0.0
Technology can increase the added value of aquaculture	0.5	0.2	0.2	0.1
Technology can stop the felling of mangroves	0.3	0.3	0.2	0.2
Technology capable of utilizing marginal lands	0.7	0.2	0.1	0.0
Technology can maintain the sustainability of the production and business	0.7	0.2	0.5	0.5

absorb ammonia and nitrite than they do oxygen, which leads to stress and mortality of shrimp.

Phytoremediation is a technology which utilizes plants to decrease or even remove contaminants within the soils and waters. Phytoremediation can be a promising method because of the low cost and its aesthetic factor which also supports environmental preservation efforts. Therefore, to anticipate developmental activity in the water body especially in aquaculture tanks, phytoremediation technology is done by utilizing plants that have the ability to store and accumulate within the their cells (phytoreaction) and the ability to metabolize (phytodegradation) contaminants for energy production and growth. One of the plant candidates to be used as a phytoremediation agent is *Gracilaria* sp.

Gracilaria sp. is also called *nitrogen starved* for its high capability of accumulating nitrogen. In addition, it is also capable of metabolizing organic contaminants as its source of nutrients for energy production and growth (Said 2005). Moreover, it has been explained that elimination of nitrogen by 500 g of *Gracilaria* sp. shows that the average nutrient concentration of nitrogen is stored by seaweeds inside their cells called thallus. This mechanism shows that integration of seaweed into an ecosystem is useful for phytoremediation of organic matters as a result of feed accumulation. This technology is targeted as an alternative technology to increase water quality.

The processing of water through biofiltration can be done aerobically, anaerobically or a combination of both aerobic-anaerobic processes (Supangat 2012). Aerobic process occurs when dissolved oxygen does exist the water reactor, whereas the anaerobic process occurs in the absence of oxygen; the aerobic-anaerobic processes is a combination of both conditions. The aforementioned process is normally utilized to remove nitrogen content in the water. Under aerobic conditions, nitrification takes place in which nitrogen ammonia is converted into nitrate. Nitrate is then converted into nitrogen gas through denitrification under anaerobic conditions. *Gracilaria* sp. can inhibit the decline of water quality in tanks during cultivation and help increase productivity of shrimp with density level of *Gracilaria* sp. that negatively correlates with concentration of ammonia (Said 2005). Absorption rate of ammonium reached 90–95%.

Field tests revealed that the model shrimp farming technology combined with recirculation systems using *Gracilaria* sp. as phytoremediation can improve water quality and shrimp growth. Maintenance of water quality was demonstrated for a relatively normal range for shrimp growth, although there were no water changes during the maintenance period. *L. vannamei* shrimp daily growth rate shows a value above the average growth in the range of 0.1–0.6 g/day. This is possibly because there is a strong linkage between the water quality and the growth of shrimp. Good water quality is required for the optimum growth and survival of shrimp. The only problem in this model is if the seaweed die, it will increase the amount of organic material in the water. The replacement or stitching of dead seaweed is required to keep the water quality stable for shrimp to grow.

In general, water quality was in the normal condition except for ammonia and organic substances which were around 8.9 mg/L and 145.6 mg/L, respectively. The data of water quality during research is shown in Table 12.3.

At the termination of the research, FCR was 1.1, and the average of survival rate was 76.7%. While weight of the biomass obtained was 59.2 kg in size of 58.3 shrimp/kg. The best biomass, size, ABW, FCR and survival rate were 66.16 g, 53 ind.kg^{-1}, 18.9 g, 0.87 and 83%, respectively. The three first results were found in tank A, and the rest were showed by tank B and C. Meanwhile, the worse results of biomass, size, ABW, FCR and survival rate were showed by tank B of about 54.3 g, tank C of about 63, tank C of about 15.9 g, tank C of about 1.27 and tank B of about 71%, respectively (Table 12.4).

Based on a simple calculation of financial analysis, the system obtained a profit of 12,000,000 IDR per cycle, with the value of B/C ratio being 1.99 and payback period at 2.01 years (Table 12.5).

The performance of the production technology model is relatively good when compared to shrimp culture activities in ponds. However, several advantages can be gained by implementing this technology:

Table 12.3 Water quality measurement results during research

No	Parameter	Result	SNI 01–7246-2006[a]
1	Temperature (°C)	20–30	28.5–31.5
2	pH	7.0–8.3	7.5–8.5
3	Salinity (g/L)	15–25	15–25
4	DO (mg/L)	4–10	>3.5
5	Ammonia (mg/L)	0.01–0.5	<0.01
6	Nitrate (mg/L)	5.24–20.48	
7	Nitrite (mg/L)	<0.01–0.5	<0.01
8	TOM (mg/L)	40–50	<55

Note: [a]Indonesian national standards for the production of shrimp *L. vannamei* with intensive technology

Table 12.4 Production performance of each system

Tank	Biomass (g)	Size	Feed (kg)	Average body weight (g)	Feed conversion ratio	Survival rate (%)
A	64.16	53	62.84	18.9	1.02	76
B	54.3	59	62.39	16.9	0.87	71
C	59	63	46.42	15.9	1.27	83
Sum	177.46	175	171.65	51.7	3.16	230
Average	59.2	58.3	57.2	17.2	1.1	76.7

Table 12.5 Financial analysis of the research model technology

Type of cost	Value (IDR)
Investment cost	21,700,000
Variable cost per cycle	3,494,332
Variable cost per year	17,471,660
Fix cost per year	4,412,000
Operational cost	21,883,660
Income per cycle	12,2 47,200
Income per year	61,236,000
Profit per cycle	8,752,868
Profit per year	43,764,340
Benefit cost ratio (%)	1.99
Payback payment	2.01

1. Environmentally friendly: this technology can reduce the level of organic pollutants in the water from culture activity, thereby supporting sustainable production and enterprise.
2. Conservation of resources: with this technology, people do not need to cut mangroves, so mangrove forests can be preserved, and people no longer need to carry out excavation for culture activity.
3. Social aspects: this technology creates new entrepreneurial opportunities in the field of household-scale shrimp culture and ultimately will reduce level of unemployment within community. Besides, the simplicity of this model is easy to be applied and accepted by society.
4. Economics: this technology gives enough benefit and is affordable to the middle-class people because it has low capital investment and operating cost.
5. Utilization of resources: this technology can be applied to marginal lands to become a productive enterprise activity.

Based on the results regarding public perception, as explained above, it is shown that 61% of respondents absolutely agree with the application of this technology as an alternative business to increase the income of low-revenue society, to reduce pollution and mangrove logging, and to support the sustainability of production activities. The development of this technology model requires both national and local government to support, especially in the infrastructure preparation and human resources as technology assistance in the society. Based on the result of interview, the majority of the respondents expressed their hope for more community empowerment activities instead of just training and seminars in order to raise their wages. They also requested for developing more activities among the people which would be supervised by the experts.

12.5 Conclusion

In conclusion, *Gracilaria* sp. is a potential candidate as phytoremediation medium for intensive shrimp farming waste. In addition, this technology is suitable to be applied in the community with environmental, economical and ease of application benefits as considerations. The growth and survival rate of *L. vannamei* and seaweed was relatively good using this technology.

References

Alex M, Theresa AR (2000) Environmental management for aquaculture. Kluwer Academic Publisher, Dordrecht

Dahuri R, Rais J, Ginting SP, Sitepu MJ (2001) Management of integrated coastal and marine area. Jakarta, PT. Pradnya Paramita

Directorate General of Aquaculture. Masterplan Brackish Aquaculture Development in Indonesia. Jakarta: Directorate General of Aquaculture; 2003, ix+164

FAO 1991 Reducing environmental impacts of coastal aquaculture. Report and Studies No. 47. Rome: IMO/FAO/Unesco/WMO/WHO/IAEA/UN/UNEP Joint GESAMP; v+35

Lin CK, Ruamthaveesub P, Wanuchsoontorn P (1993) Integrated culture of the green mussel (*Perna viridis*) in wastewater from an intensive shrimp pond: concept and practice. World Aquacult 24:68–73

Pillay TVR (2004) Aquaculture and the environment, 2nd edn. Blackwell Publising, Garsington Road. xii+196

Said NI, Ruliasih (2005) Overview technical aspects of the electoral media biofilter for wastewater treatment. J Aquacult Indonesia 1(3):1–10

Supangat A (2012) Biofilter system to improve water quality shrimp. J JBPTUNIKOMPP

Troell M, Ronnback P, Halling C, Kautsky N, Buschmann AH (1999) Ecological engineering in aquaculture: use of seaweeds for removing nutrients from intensive Mariculture. J Appl Pycol 11:89–97

Part V
Food Security and Technology

Chapter 13
Effect of Salt Stress on Growth Performance and Antioxidant Status of Local Black Rice (*Oryza sativa L.*)

Yuda C. Hariadi, Arry Y. Nurhayati, and Rida R. Akbar

Abstract *Oryza sativa* L. or black rice is a well-known source of carbohydrate for baby food and diet programs because of its high nutrient content, but it is rarely cultivated in Indonesia. We aimed to investigate how well the species tolerates salt stress and how salt affects its antioxidant status. Ion leakage from the root tissue was observed based on the salinity level of the growth medium to select salt-tolerant plants. The results show that this is a viable method for screening salt-tolerant rice. Our study results could promote the use of black rice to prevent cardiovascular diseases and cancer while ensuring sustainable cultivation of the local best rice species in the future, as well as improved economic return for farmers.

Nomenclature

MIFE	Microelectrode ion flux estimation
PD	Potential difference
ANOVA	Analysis of variance
EC	Electrical conductivity
TDS	Total dissolved salt
ORP	Oxidation–reduction potential
rH	Absolute indicator of reductive potential, indicator of hydrogen content

Y.C. Hariadi (✉) • A.Y. Nurhayati
Physics Department Faculty of Mathematics and Natural Sciences, University of Jember, Jember 68121, East Java, Indonesia
e-mail: yuda_ch@utas.edu.au

R.R. Akbar
Pharmacy Faculty, University of Jember, Jember 68121, East Java, Indonesia

B. McLellan (ed.), *Sustainable Future for Human Security*,
https://doi.org/10.1007/978-981-10-5430-3_13

13.1 Introduction

Reduction of rice yield due to climate change and an increase in natural disasters has affected Indonesia and other developing countries. Rice has been an important primary cereal for generations (Padhye and Salunka 1979) with 90% consumption and production in Asia (FAO 2008). Extensive efforts have been made to increase rice productivity in Indonesia, to fulfill the nutritional demand of three million babies born annually, and to reduce dependence on import. This need for high rice yields has resulted in some of the best rice cultivars becoming scarce and being replaced with fast-growing and high productivity rice that might have inferior taste, fragrance, and nutritional status.

Black rice has been known for decades, but because of its long growing period, it is not cultivated much. This was a rice species owned by farmers, usually planted for baby food or as a diet food. Black rice has been shown to have a high protein content (8.16%) and a low fat content (0.07%) (Thomas et al. 2013), while basmati has a protein content ranging from 7.32 to 7.78% (Ahmad et al. 1985). Choosing to cultivate red or black rice can improve the quality of the best rice species and strengthen sustainability by preserving the local black rice species. In addition, the cultivation of local black rice could preserve the natural resources beneficial for future generations. Considering the increasing population, the cultivation of better rice species can be highly beneficial for securing food and economic sustainability, as well as increased community research and free seed exchange, which might be limited in developing countries because of property rights (Jacobsen et al. 2013).

We aimed to investigate the performance of local black rice and compare its nutritional status with that of colorless rice. Several methods have been previously used to assess adaptability to salt, including microelectrode ion flux estimation (MIFE) on quinoa (Hariadi et al. 2011) and leaf surface electrical measurement in screening of rice (Hariadi et al. 2015). Here, we measured leakage of ions from the root tissue. This method could indicate salt tolerance of plants during early germination. Further, integrating this method with saline remediation, as well as its nutritional status, will help promote consumer awareness of the health benefits of black rice grains (Sompong et al. 2011) and pigment (Wang et al. 2007), which imparts positive nutritional properties such as prevention of cardiovascular diseases and cancer. The anthocyanin pigment content and antioxidative and anti-inflammatory properties of black rice might have potential uses in nutraceutical or functional food formulation (Hu et al. 2003), as well as in antiaging cosmetic materials and health food (Kaneda et al. 2006).

13.2 Material and Methods

13.2.1 Plant Material

The black rice seeds were obtained from farmers in Jember, East Java, Indonesia. Homogeneous seeds were perfused with 0.5% sodium hypochlorite before germinating in sterilized media containing modified Hoagland solution (Hariadi and Shabala 2004). After 3 days, the survival rate was calculated. Healthy and homogenous seedlings were carefully selected and continued to be grown in media with different levels of salinity (0 [control], 50, 100, 150, and 250 mM NaCl) and sufficiently watered daily.

13.2.2 Electrophysiology and Physiology Measurements

Measurement of membrane potential difference (PD) was done by impaling a microelectrode (tip size of about 2 μm and back-filled with 1 M KCl) into the black rice epidermal root cells of 3-day-old rice seedlings in a Faraday cage under controlled room temperature of 27 ± 2 °C. A chamber filled with basic salt medium (BSM containing 0.1 mM $CaCl_2$ and 0.2 mM KCl) (pH 5.6 unbuffered) was used for this measurement. The electrode was manually directed to the target cell by using a micromanipulator (Fig. 13.1). Once a steady membrane PD was observed, perfusion of media containing different levels of salt, 0, 50, 100, 150, and 250 mM NaCl, was performed. The membrane PD was measured (ion leakage from the root tissue) continuously by using a four-channel electrometer (Department of Physics University of Tasmania, Australia) and recorded on a computer. The growth and development of black rice were measured and compared with those of the control plant.

13.2.3 Antioxidant Status

Antioxidant status was investigated by determining the rH, before measuring the total protein and phenolic contents of fresh roots and shoots of plants grown under different salinity. The antioxidant status, indicated by the oxidation–reduction potential (ORP), and pH were assessed to obtain the rH value. The rH level was a criterion of the state of reduction or oxidation and indicator of the probability that the compound will react with the free radical. Here, we used the Polisǎk method (2008). The antioxidant status was measured at 25 °C to meet with the equation.

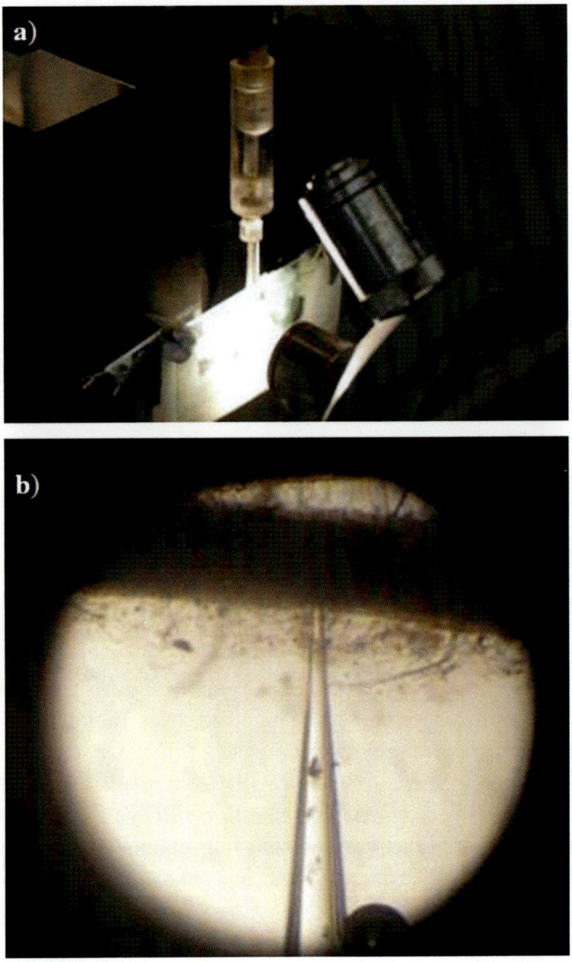

Fig. 13.1 Membrane potential difference (PD) measured in rice root epidermal cells. Preparing seedling in a chamber: (**a**) chamber clip of the microscope and (**b**) impaling the microelectrode into the epidermal rice root cell (viewed at 100×)

13.3 Results and Discussion

13.3.1 Electrophysiology and Physiology Performance

The adaptability of rice under adverse conditions or sensitivity during germination might have a significant effect on yield reduction (Jacobsen, n.d.), i.e., low seed quality or dense sowing might affect rice survival rate. In our experiments, the germination rate was 100%, showing good quality of the seeds from the farmers. Since rice is an integral part of Asian culture and has been planted for many centuries, data on rice have been well documented for breeder and grower, but not for the variety grown, such as black rice, red rice, or purple rice. In addition, the seeds were not readily available unlike those of white rice, at least in East Java, Indonesia. Although all seedlings grew well, the rate of growth differed slightly

Fig. 13.2 Performance of seedling growth of black rice (**a**) compared to (**b**) white rice, under different salinity levels of 250 mM, 150 mM, 100 mM, 50 mM, and control (from *left to right*)

between rice seedlings grown under sufficient light compared to those grown under darker and warmer conditions such as in the growth chamber.

Observations on the second day after exposure to different salinity levels in the growth medium showed that compared to the control, 100 mM or higher NaCl levels affected seedling growth with reduction in seedling height or shoot length (Fig. 13.2). Seedling adaptability to 50 mM NaCl was higher for black rice compared to white rice on the second day of exposure but lower on the third day. The adaptability was relatively similar to that of white rice on the second and third day for salinity levels higher than 100 mM NaCl.

The ratio of shoot to root length tended to increase after decreasing for black rice, while for white rice, the same ratio was maintained on the second day but decreased on the third day (Fig. 13.3). The seedling and plant heights of black rice were lower than those of white rice, but the ratio of shoot to root length did not differ much between control black and white rice on the second day, at 1.87 and 1.96, respectively. Increasing the salinity to 250 mM NaCl resulted in an increase in the shoot-to-root length ratio to 2.76 for black rice but decreased the ratio to 1.70 for white rice.

The fresh and dry weight ratios were measured without the grain still attached to the plants. The grain was heavier than the shoot or the root, and black rice can be differentiated from white rice by the color of the grain husk, but the seed itself was of a color similar to other rice, which changed only after being immersed in water (Figs. 13.4 and 13.5).

The shoot-to-root fresh weight ratio of black rice on the first day and third day of salinity exposure ranged from 0.74 to 1.20 and was lower than that of white rice, which ranged from 1.30 to 1.27 and decreased on the second day but increased on the third day (Fig. 13.6). The ratio decreased for 50 mM NaCl on the first day exposure and increased the following day, contrary to the effect in white rice, where the ratio increased on the first day and decreased the following day. The behavior of black rice differed under different salinity levels, except for the increase in shoot-

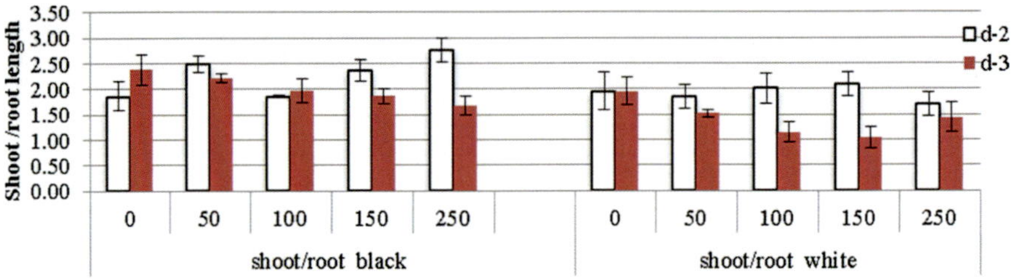

Fig. 13.3 Shoot-to-root length ratio of black rice and white rice grown under different salinity levels on days 2 and 3 of measurement

Fig. 13.4 (**a**) Germination of different-day-old white and black rice (*Oryza sativa*) and (**b**) black rice seedlings

to-root fresh weight ratio on the third day. In white rice, except for the 150 mM NaCl, the shoot-to-root fresh weight ratio decreased, while it remained the same in the control.

Although the height of black rice was less than that of white rice (Fig. 13.5), the shoot-to-root dry weight ratios were higher for the control and 50 mM NaCl (Fig. 13.7). The comparison of root growths under different salinity levels is shown in Fig. 13.6. Shoot-to-root fresh weight ratio ranged 0.69 to 1.36 for control black rice and 0.83 to 1.83 for control white rice. In black rice, under 50 mM and 150 mM NaCl, the ratio increased to 1.61 on the first day; under 250 mM NaCl, the ratio decreased on the second and third days. Thus, higher salinity tended to reduce the shoot-to-root dry weight ratio for white rice (Fig. 13.7).

Previous studies (Hariadi et al. 2015) have shown that rice adaptability to salinity was similar to that of quinoa (Hariadi et al. 2011), but since Indonesia has a high diversity of rice plants, rice could be grown under higher salinity. For quinoa, optimal growth was achieved at approximately 100–200 mM NaCl[6] but could be adapted to a higher level of approximately 400 mM NaCl, with not much difference in performance when grown in soil with high salinity in the beginning or

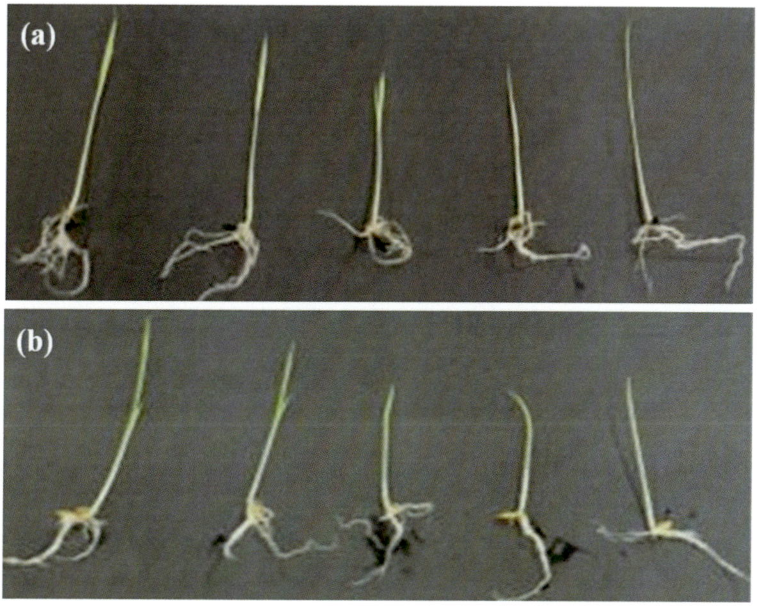

Fig. 13.5 (**a**) Roots of black rice seedlings grown under different salinity levels. (**b**) Roots of white rice grown under different salinity levels of 50 mM, 100 mM, 150 mM, and 250 mM NaCl (from left [control] to right)

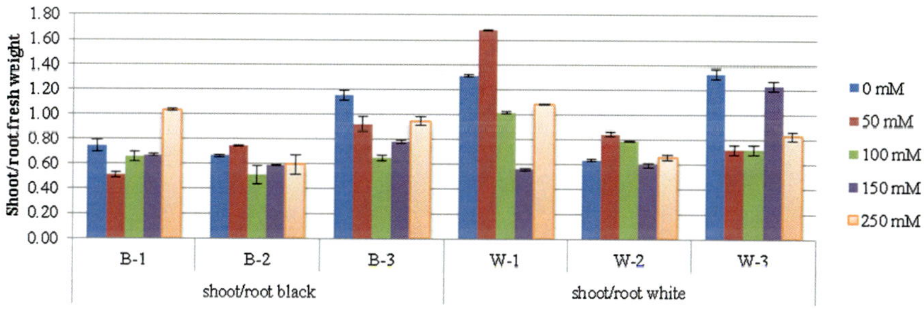

Fig. 13.6 Shoot-to-root fresh weight ratio of black rice (B) and white rice (W) grown under different salinity levels

when grown by gradual additions of NaCl[6]. However, cultivation of rice at 200 mM NaCl for short durations was not possible.

Saline stress inhibited the germination of black rice by reducing seedling height, fresh weight, and relative dry weight. Our results show that seed germination is an important stage in salinity adaptation process and key to survival of rice on the growth medium. The growth of black rice was affected by the different salinity levels. The difference in tolerance to salinity was possibly due to ionic relation involved in the salt tolerance mechanism in the root of the plant. Salt tolerance in

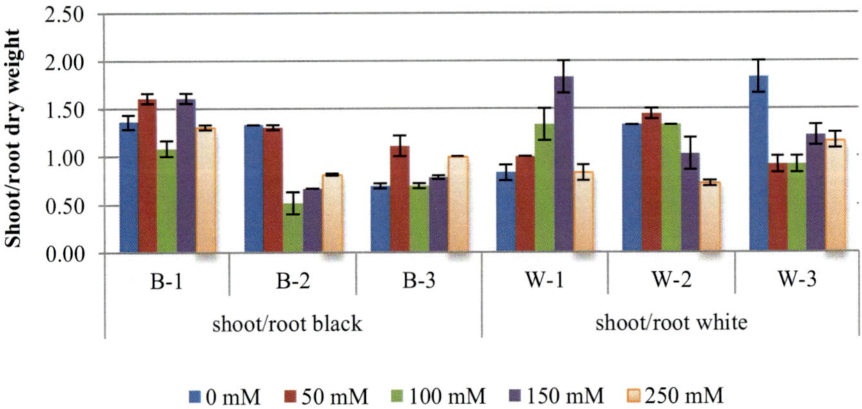

Fig. 13.7 Shoot-to-root dry weight ratio of black rice (B) and white rice (W) grown under different salinity levels

the root during the early seedling stage will support adaptability to salt stress in the later stages of plant growth.

Black rice growth is affected by salinity of the growth medium from germination to early stages of vegetative growth. Our results support the results of a previous study in that the growth of rice was affected from the germination stage and growth (Hariadi et al. 2015) and can be used for screening process. Further research is needed to improve the performance of black rice and for better management to reduce the cost due to the long harvesting time required compared to other colorless rice. Monitoring the effects of salt stress on different rice varieties is also important for remediation process of saline environments, especially during the seed germination stage, which is critical to ensure good productivity (Dash 2012).

13.3.2 Membrane Potential Difference

The electrophysiological performance of black rice can be analyzed from the membrane potential difference (PD) during depolarization and repolarization in the root cell. The release of ions such as K^+, Ca^{2+}, and H^+ can be considered as depolarization (Chen et al. 2005). K^+ comprises the majority of ions leaked (flux of ions across the cell membrane) causing the depolarization (Hariadi et al. 2011; Chen et al. 2007). High depolarization of the root cell membrane at certain salinity levels means the rice plant is less tolerant of that salinity condition (Figs. 13.8 and 13.9). The trend of depolarization in both graphs showed a dose-dependent response to the different salinity levels. Increasing amplitude meant increased sensitivity of rice to the salinity stress.

In black rice, 250 mM NaCl resulted in higher depolarization, and thus, we can conclude that black rice is more sensitive than white rice. However, both rice types

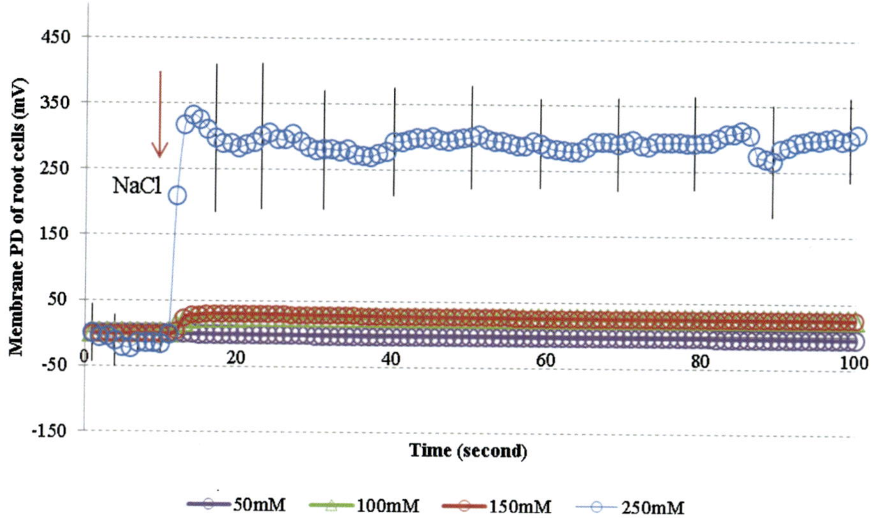

Fig. 13.8 Transient membrane potential difference (PD) in root cells of black rice plants ($n = 4$) grown under different salinity levels (applied as indicated by the arrow), means ± standard error

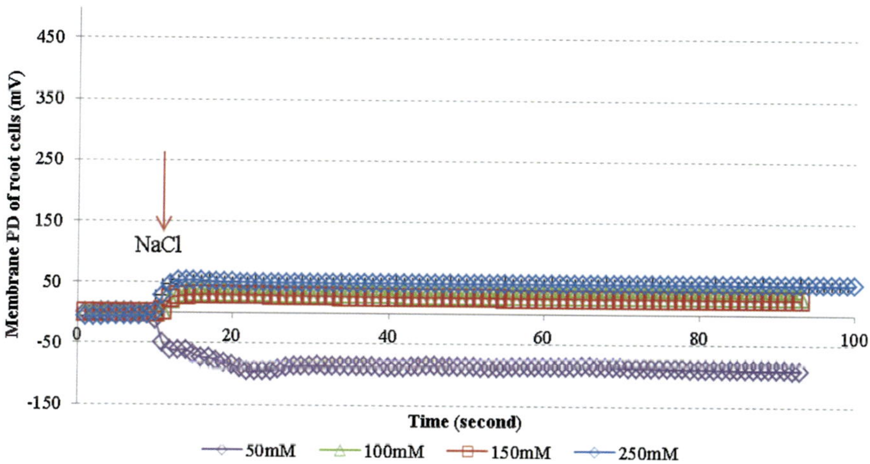

Fig. 13.9 Transient membrane potential difference (PD) in root cells of white rice plants ($n = 4$) grown under different salinity levels (applied as indicated by the arrow), means ± standard error

showed tolerance to 150 mM NaCl. Thus, membrane PD can be used as an indicator of the performance of rice under salt stress. Further, leakage of K^+ is considered the best indicator of whether plants are sensitive or tolerant to salt stress (Chen et al. 2007).

In our results, the initial resting membrane PD of root cells was normalized to understand the transient membrane PD after perfusion with different NaCl levels. The amplitude of the different NaCl levels was related to the performance of rice to

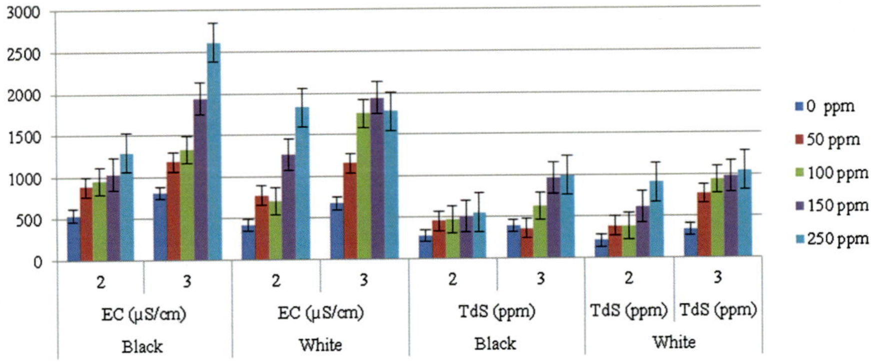

Fig. 13.10 Electrical conductivity (EC) and total dissolved salt (TDS) of black rice compared to white rice grown under different salinity levels of 0, 50, 100, 150, and 250 mM

the respective salinity level. Compared to MIFE, measuring membrane potential as a performance indicator for salt tolerance in rice varieties is easier and cheaper and saves more time (Hariadi et al. 2015), but MIFE is necessary to identify the ions responsible for depolarization and repolarization causing the transient membrane PD in rice root cells under different salinity levels.

13.3.3 Antioxidant Status

The electrical conductivity and TDS in seedling rice increased with increase in the salinity (Fig. 13.10) for both black and white rice. At 250 mM NaCl, the electrical conductivity increased from the normal value of approximately 750 µS/cm to approximately 2500 µS/cm on the third day.

The rice antioxidant status of control black rice seemed greater than that of white rice, as indicated by the greater rH value. The rH value of both black and white rice increased under 50 mM NaCl on the second day of exposure but decreased on the third day for black rice and increased for white rice; the rH at 100 mM and 150 mM NaCl decreased in white rice (Fig. 13.11), contrary to that in black rice, in which the antioxidant status was high on the third day.

The effect of 250 mM NaCl on rH was similar in black and white rice, increasing on the second day and then decreasing on the third day. Figure 13.12 shows that the antioxidant level tended to be higher for black rice than for white rice on the third day, with increasing rH for 100 and 150 mM NaCl. It is possible that compared to white rice, black rice contains higher proteins and phenolics, but further research is needed to calculate the values.

Flavonoids are one of the main phenolic components in diet. Consuming black rice could lower cardiovascular risk factors in patients (Wang et al. 2007). Although rice does not have a high concentration of phenolic compounds (Walter and Marchesan 2011), increased consumption of black rice could mitigate the need for a cereal

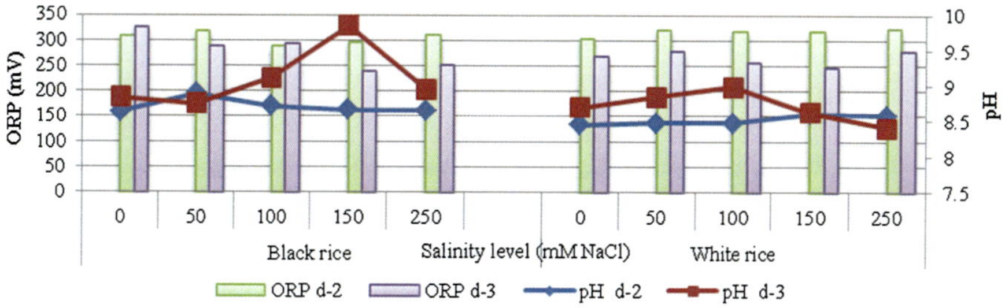

Fig. 13.11 Oxidation–reduction potential (ORP) and pH of black rice and white rice grown under different salinity levels (0, 50, 100, 150, and 250 mM)

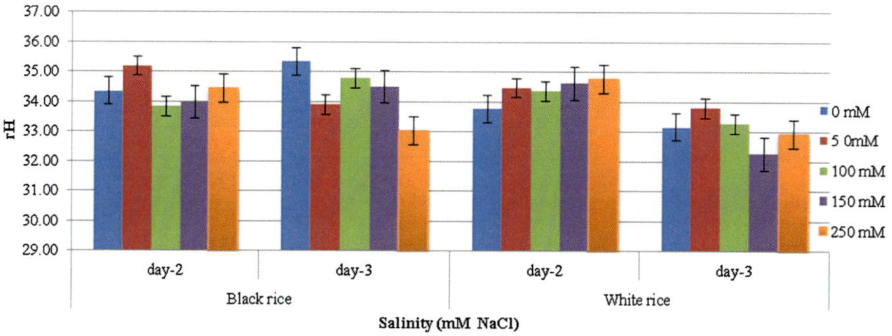

Fig. 13.12 Comparison between antioxidant levels of black rice grown under different salinity levels (0, 50, 100, 150, and 250 mM) with those of white rice

with high nutritional status, and better prices would increase farmer income. The use of black rice with high nutritional quality as a healthy and vegetarian diet and in baby food will also increase its market value, which will encourage more farmers to cultivate this rice, thereby increasing the productivity. Black rice is similar to quinoa in terms of its behavior under saline conditions and as a beneficial food. In quinoa, some phenomenon underlying salt tolerance has been reviewed (Adolf et al. 2013). Using the knowledge gained will ensure better adaptation process and performance.

13.4 Conclusion

Compared to white rice, black rice was more sensitive to 250 mM NaCl; however, both rice plants showed similar electrophysiological tolerance at 150 mM NaCl. The measurement of membrane potential as performance indicator of salinity tolerance is easy and cheap and saves time unlike measurement methods such as MIFE. However, using MIFE could help identify the ions responsible for depolarization and repolarization causing transient membrane PD in rice root cells under different salinity levels.

The different effects of salinity on growth and nutritional status of black rice require further research to explore the performance of black rice. Our results suggest the potential use of black rice as high-quality food and its cultivation for promoting environmental quality and sustainability and for protecting the local best species for future generations and better economic return for farmers.

The nutritional status of black rice and its adaptability to salt stress could promote rice planting in coastal areas, which in turn could increase consumption of black rice and reduce dependence on rice import.

Acknowledgments The author wishes to thank Dr. Ian Newman, Department of Physics University of Tasmania, Australia, for providing the electrometer in the experiment.

References

Adolf VI, Jacobsen S-E, Shabala S (2013) Salt tolerance mechanism in quinoa (*Chenopodium quinoa* wild). Environ Exp Bot 92:43–54

Ahmad MR, Khan DM, Hamdard MS (1985) Nutritive value of different varieties of rice. Pak J Agric Res 6(1):8–11

Chen Z, Newman I, Zhou N, Mendham N, Zhang G, Shabala S (2005) Screening plants for salt tolerance by measuring K^+ flux: a case study for barley. Plant Cell Environ 28:1230–1246

Chen Z, Pottosin II, Cuin TA, Fuglsang AT, Tester M, Jha D, Zepeda-Jazo I, Zhou M, Palmgren MIG, Newman I, Shabala S (2007) Root plasma membrane transporters controlling K^+/Na^+ homeostasis in salt stressed barley. Plant Physiol 145(4):1714–1725

Dash AK (2012) Impact of domestic waste water on seed germination and physiological parameters of rice and wheat. Int J Recent Res Appl Stud 12(2):280–286

FAO (2008) Food outlook global market analysis, pp 23–28

Hariadi Y, Shabala S (2004) Screening broad beans (*Vicia faba*) for magnesium deficiency II. Photosynthetic performance and leaf bioelectrical responses. Funct Plant Biol 31 (5):539–549

Hariadi YC, Marandon K, Tian Y, Jacobsen S-E, Shabala S (2011) Ionic and osmotic relations in quinoa (*Chenopodium quinoa* wild) plant grown at various salinity levels. J Exp Bot 62 (1):185–193

Hariadi YC, Nurhayati AY, Soeparjono S, Arif I (2015) Screening six varieties of rice (*Oryza sativa*) for salinity tolerance. Procedia Environ Sci 28:78–87

Hu C, Zawistowski J, Ling W, Kitts DD (2003) Black rice (*Oryza sativa* L. Indica) pigmented fraction suppresses both reactive oxygen species and nitric oxide in chemical and biological model systems. J Agric Food Chem 51:5271–5277

Jacobsen S-E (2003) The worldwide potential for quinoa (*Chenopodium quinoa* wild). Food Rev Int 19(1 and 2):167

Jacobsen SE, Sorensen M, Pedersen SM, Weiner J (2013) Feeding the world: genetically modified crops versus agricultural biodiversity. Agron Sustain Dev 33(4):651–662

Kaneda I, Kubo F, Sakurai H (2006) Antioxidative compounds in the extracts of black rice brans. J Health Sci 52(5):495–511

Padhye VW, Salunka DK (1979) Extraction and characterization of rice proteins. Cereal Chem 56 (5):389–393

Polisăk B (2008) The amount of active hydrogen in the selected juices and teas on the Slovenian market. Int J Sanitary Eng Res 2(1):58–65

Sompong R, Sibenhandl-Ehn S, Linsberer-Martin G, Berghofer E (2011) Physicochemical and antioxidative properties of red and black rice varieties from Thailand, China and Sri Lanka. Food Chem 124:132–140

Thomas R, Wan-Nadiah WA, Bhat R (2013) Physiochemical properties, proximate composition, and cooking qualities of locally grown and imported rice varieties marketed in Penang, Malaysia. Int Food Res J 20(3):1345–1351

Walter M, Marchesan E (2011) Phenolic compounds and antioxidant activity of rice. Braz Arch Biol Technol 54(1):371–377

Wang Q, Han P, Zhang M, Xia M, Zhu H, Ma J, Hou M-J, Tang Z, Ling W (2007) Supplementation of black rice pigment fraction improves antioxidant and anti-inflammatory status in patients with coronary heart disease. Asia Pac J Clin Nutr 16(Suppl 1):295–301

Chapter 14
Larvicidal Activity of *Cerbera odollam* Gaertn Against a Dengue Vector, *Aedes aegypti* (Diptera: Culicidae)

Didi Tarmadi, Dwi Jayanti Gunandini, and Sulaeman Yusuf

Abstract Mosquitoes act as vectors of hazardous human diseases and are a major public health problem. *Aedes aegypti* is a vector of dengue. Chemical treatment is an effective method of controlling mosquitoes but is unsafe for the environment. It is crucial that new, environmentally safe products be developed. Plants can be used as an alternative source of mosquito larval control materials. This study was conducted to evaluate the larvicidal activity of *Cerbera odollam* against a dengue vector, *A. aegypti*. The larvicidal activity of crude extracts, fractions, and subfractions of *C. odollam* was tested against the third-instar larvae of *A. aegypti*. The larval mortality was determined at 6, 12, 24, and 48 h of exposure. The results showed that the crude extracts of rinds, stem bark, leaves, and seed kernels had poor larvicidal effects. However, the highest larval mortality was shown when the crude extract of seed kernels was used at a concentration of 2.0 g/l after an exposure of 48 h. The larvicidal activity of the fractions derived from the crude extract of seed kernels led to high larval mortality in the ethyl acetate fraction, moderate mortality in the n-hexane fraction, and a very low larvicidal effect in the insoluble fraction. The LC_{50} values of the ethyl acetate and n-hexane fractions were 0.07 and 0.14 g/l, whereas the LC_{90} values were 0.42 and > 1.0 g/l, respectively. The larvicidal activity

D. Tarmadi (✉)
Research Center for Biomaterials, Indonesian Institute of Sciences (LIPI), Jl. Raya Bogor KM. 46, Cibinong, Bogor 16911, Indonesia

Research Institute for Sustainable Humanosphere (RISH), Kyoto University, Gokasho, Uji, Kyoto 611-0011, Japan
e-mail: didi@biomaterial.lipi.go.id

D.J. Gunandini
Faculty of Veterinary Medicine, Bogor Agricultural University, Jl. Agatis, Kampus IPB Darmaga, Bogor 16680, Indonesia
e-mail: djg_0205@yahoo.com

S. Yusuf
Research Center for Biomaterials, Indonesian Institute of Sciences (LIPI), Jl. Raya Bogor KM. 46, Cibinong, Bogor 16911, Indonesia
e-mail: Sulaeman@biomaterial.lipi.go.id

B. McLellan (ed.), *Sustainable Future for Human Security*,
https://doi.org/10.1007/978-981-10-5430-3_14

of the subfractions yielded the highest larval mortality in subfractions 1 and 7. The bioactive compounds of *C. odollam* possess potency as larvicidal substances.

14.1 Introduction

Mosquitoes serve as vectors of several diseases that cause serious health problems in humans (Markouk et al. 2000). *Aedes aegypti*, a vector of dengue, is widely distributed in tropical and subtropical zones (Rahuman et al. 2008a). Dengue is primarily an urban disease of the tropics and is a hazardous disease that can be fatal. Dengue is caused by one of four virus serotypes (DEN-1, DEN-2, DEN- 3, and DEN-4). Infection with one of these serotypes does not reserve cross-protective immunity, and humans living in a dengue endemic district are able to have four dengue infections over their lifetimes (Gubler and Clark 1995). One of the efforts to prevent dengue dispersion involves the management of the vector (World Health Organization 2005). Chemical treatment is an effective method used extensively in daily life. Several insecticides are effective including organochlorine, organophosphorus, carbamates, pyrethroids, and others. Nevertheless, the control of mosquitoes with insecticides has caused environmental problems, and the resistance of mosquitoes to insecticides has increased during the last five decades (Govindarajan and Karuppannan 2011). In addition, it was reported that the constant use of chemical insecticides has often led to the disruption of natural biological control systems, outbreaks of insect species, and undesirable effects on humans, mammals, and other nontarget organisms (Brown 1986; Chaithong et al. 2006). Researchers are developing new products for the management of mosquitoes that are safer for the environment, biodegradable, low in cost, reliable, and functional for preventive and remedial management.

Plants can be used as an alternative source of mosquito larva management agents because they are a rich source of bioactive chemicals (Park et al. 2002). Earlier researchers reported that several plant extracts possess potency as larvicidal substances against a dengue vector, *A. aegypti*; these substances include the dichloromethane extracts of *Abuta grandifolia* and *Minthostachys setosa* (Ciccia et al. 2000); the methanol fruits extract of *Piper longum* (Yang et al. 2002); the fruits extract of *Piper nigrum* (Park et al. 2002); the saponin extract of *Quillaja saponaria* (Pelah et al. 2002); the leaf and bark essential oil of *Cryptomeria japonica* (Cheng et al. 2003); the ethanol seed extract of celery, *Apium graveolens* (Choochote et al. 2004); the acetone and petroleum ether extracts of *Murraya koenigii*, *Coriandrum sativum*, *Ferula asafoetida*, and *Trigonella foenum gracium* (Harve and Kamath 2004); the ethanolic extract of *Azadirachta indica* (Wandscheer et al. 2004); and the ethyl acetate, butanol, and petroleum ether extracts of *Jatropha curcas*, *Pedilanthus tithymaloides*, *Phyllanthus amarus*, *Euphorbia hirta*, and *Euphorbia tirucalli* (Rahuman et al. 2008a).

Several essential oils derived from plants exhibited larvicidal activity against *A. aegypti* including essential oils of *Zingiber officinale*, *Rosmarinus officinalis*, and *Cinnamomum zeylanicum* (Prajapati et al. 2005); *Anacardium occidentalis*, *Copaifera langsdorffii*, *Carapa guianensis*, *Cymbopogon winterianus*, and *Ageratum conyzoides* (de Mendonça et al. 2005); *Tagetes patula* (Dharmagadda et al. 2005); *Croton nepetaefolius*, *Croton argyrophyloides*, *Croton sonderianus*, and *Croton zehntneri* (Morais et al. 2006); *Mammea siamensis*, *Anethum graveolens*, and *Annona muricata* (Promsiri et al. 2006); and *Chloroxylon swietenia* (Kiran et al. 2006).

Ethanolic extracts of some Brazilian medicinal plants showed larvicidal activity against *A. aegypti* (De Omena et al. 2007). Leaf extract of *Melia azedarach* (Coria et al. 2008) and the leaf extracts of *Ocimum canum*, *Ocimum sanctum*, and *Rhinacanthus nasutus* exhibited larvicidal activity against the early fourth-instar larvae of *Spodoptera litura*, *A. aegypti*, and *Culex quinquefasciatus* (Kamaraj et al. 2008). The acetone, chloroform, ethyl acetate, hexane, and methanol leaf extracts of *Acalypha indica*, *Achyranthes aspera*, *Leucas aspera*, *Morinda tinctoria*, and *Ocimum sanctum* were tested against *A. aegypti* and *Culex quinquefasciatus* (Bagavan et al. 2008). The leaf extracts of *Leucas aspera* (Maheswaran et al. 2008), the saponin from *Balanites aegyptiaca* (Chapagain et al. 2008), the leaf extract of *Millingtonia hortensis* (Kaushik and Saini 2008), the ethanol extract of the leaves and flowers of *Lantana camara* (Kumar and Maneemegalai 2008), the benzene leaf extract of *Citrullus vulgaris* (Mullai et al. 2008), the leaf extract of *Cassia fistula* (Govindarajan 2009), the benzene and ethyl lectins isolated from *Myracrodruon urundeuva* bark (Sá et al. 2009), and the constituents of *Asarum heterotropoides* root stem showed significant larvicidal activity against the larvae of *A. aegypti* (Govindarajan et al. 2011). Similarly, acetate leaf extracts of *Ervatamia coronaria* and *Caesalpinia pulcherrima* (Perumalsamy et al. 2009); the leaf extracts of *Andrographis paniculata* (Govindarajan 2011); the benzene, hexane, ethyl acetate, methanol, and chloroform leaf extracts of *Eclipta alba* (Govindarajan and Karuppannan 2011); and the leaf extract of *Acalypha alnifolia* also showed larvicidal activity against larvae of *A. aegypti* (Kovendan et al. 2012).

This study was carried out to investigate the larvicidal activity of *Cerbera odollam* against a dengue vector, *A. aegypti*, as *C. odollam* has been known as a poisonous tree (Gaillard et al. 2004). Previous studies showed that a subterranean termite of *Coptotermes gestroi* (Tarmadi et al. 2007) and a plant pest of *Eurema* spp. showed high mortality rates when treated with the extract methanol of *C. odollam* (Utami 2010). Hashim et al. (2009) reported that methanol wood extracts from *C. odollam* showed the high activities against *Trametes versicolor*, *Pycnoporus sanguineus*, and *Schizophyllum commune*. The methanol seed extract of *C. odollam* showed moderate antibacterial activity against *Salmonella typhi*, *Streptococcus saprophyticus*, and *Streptococcus pyogenes* (Ahmed et al. 2008). Based on a literature survey, the larvicidal activity of *C. odollam* against the dengue vector *A. aegypti* has not yet been reported.

14.2 Materials and Methods

14.2.1 Plant Materials

Plant material. Leaves, rinds, stem bark, and seed kernels of *C. odollam* were collected in Bogor, Indonesia. Identification of the plant material was performed by the Herbarium Bogoriense, Indonesian Institute of Sciences (LIPI), where voucher specimens were deposited.

14.2.2 Preparation of the Extracts

Dried leaves (1 kg), stem bark (1 kg), rinds (1 kg), and seed kernels (2 kg) were powdered mechanically using a commercial blender and extracted with methanol. Dried crude extract was obtained using a rotary evaporator (RV 10 Digital, IKA Works GmbH & Co., Germany) at 40 °C. Crude extract of the seed kernels was continued for the fractionation steps. A total of 175 g of the dried crude extract of seed kernels was dissolved in 600 ml distilled water and n-hexane (1:1). Furthermore, a distilled water fraction was separated and added to 300 ml ethyl acetate for further extraction. 24 g dried extract of the ethyl acetate fraction was subjected to column chromatography (100 × 3 cm, Si gel 60–120 mesh, 350 g). Elution was carried out with different proportions of n-hexane and chloroform. Subfractions were collected by thin-layer chromatography.

14.2.3 Larvicidal Bioassay

The crude extracts were prepared in different concentrations: 0.01, 0.25, 0.5, 1.0, and 2.0 g/l. The fractions were 0.01, 0.25, 0.5, and 1.0 g/l, whereas the subfractions were prepared at the concentration of 1.0 g/l. Tween 80 was used as a surfactant. The larvicidal activity was assessed by using the procedure of WHO 2005 (World Health Organization 2005). Twenty-five third-instar *A. aegypti* larvae were subjected to 100-ml solutions of each concentration in 5 replications, and the numbers of dead larvae were counted at 6, 12, 24, and 48 h of exposure. The control was set up with water and Tween 80.

14.2.4 Insect Rearing

A. aegypti eggs were obtained from the Laboratory of Parasitology and Health Entomology, Faculty of Veterinary Medicine, Bogor Agricultural University (IPB), and reared at the Laboratory of Biodeterioration and Pest-Insect Control, Research Center for Biomaterials LIPI. The larvae were reared in plastic trays containing tap water and maintained at a temperature of 28 ± 2 °C and a relative humidity of 75–85%. The larvae were fed a diet of dog food pellets.

14.2.4.1 Statistical Analysis

The larval mortality data were subjected to probit analysis for calculating LC_{50} and LC_{90} using the software EPA Probit Analysis Program Version 1.5. Chi-square values were calculated using SPSS ver. 23 Software (IBM, Armonk, NY). Results with $p < 0.05$ were considered to be statistically significant.

14.3 Results and Discussion

14.3.1 Larvicidal Activity of Crude Extracts

The crude extracts of the rind, stem bark, leaf, and seed kernel showed low larvicidal effects. However, the highest larval mortality was found in the crude extract of seed kernels at a concentration of 2.0 g/l after 48 hours of exposure (Fig. 14.1). The LC_{50} and LC_{90} values of crude extracts of the rind, bark, and leaf were more than 1.0 g/l. On the other hand, the LC_{50} and LC_{90} values of crude extract of seed kernels were 0.76 g/l and more than 1.0 g/l (Table 14.1), respectively. It was reported that *C. odollam* is a poisonous plant belonging to the Apocynaceae family (Gaillard et al. 2004), and its seeds are extremely toxic and contain cerberin, neriifolin, and other chemical compounds (Laphookhieo et al. 2004).

Earlier researchers reported that petroleum ether crude extracts of *Jatropha curcas* and *Euphorbia tirucalli* possessed larvicidal activity against *A. aegypti* with LC_{50} and LC_{90} values of 8.79 and 35.39 ppm, respectively, for *J. curcus* and 4.25 and 13.44 ppm, respectively, for *E. tirucalli* (Rahuman et al. 2008a). The ethanol crude extracts of *Annona glabra* and *Anacardium occidentalis* exhibited LC_{50} and LC_{90} values of 27 and 64 ppm, respectively, for *A. glabra* and 14.5 and 48 ppm, respectively, for *A. occidentalis* against early fourth larvae *A. aegypti* (de Mendonça et al. 2005). The larvicidal activity of ethanol crude extracts of *Annona crassiflora*, *Annona squamosa*, *Derris* sp., *Erythrina mulungu*, and *Pterodon polygalaeflorus* showed LC_{50} values of 8.94, 5.12, 8.5, 67.9, and 35.7 ppm and LC_{90} values of 34, 21.4, 15.2, 125, and 63.3 ppm, respectively, against larvae of *A. aegypti* (De Omena et al. 2007). Several crude extracts derived from plants have reportedly possessed larvicidal activity against larvae of *A. aegypti*. These included the crude extracts of *Phyllanthus amarus* (LC_{50} 90.92 ppm, LC_{90} 384.19 ppm), *Ocimum canum* (LC_{50} 99.42 ppm, LC_{90}

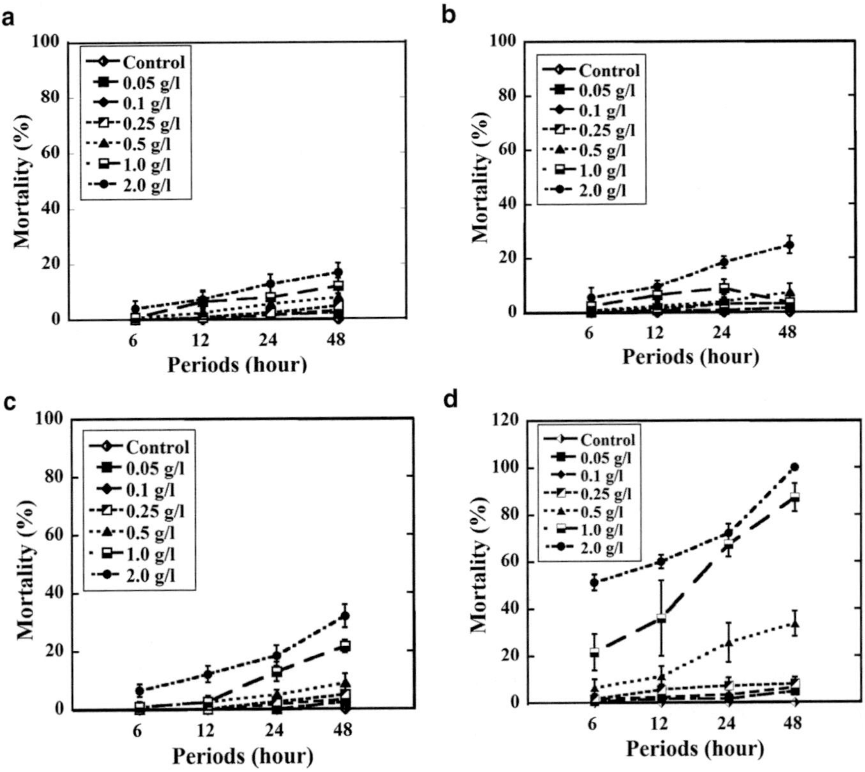

Fig. 14.1 Mortality of *A. aegypti* after exposure to crude extracts of *C. odollam*. (**a**) = stem bark, (**b**) = leaf, (**c**) = rind, (**d**) = seed kernel. Twenty-five larvae of *A. aegypti* were used per replication, and the data were averaged ($N = 5$). Error bars represent standard deviations

640.60 ppm), *Rhinacanthus nasutus* leaf (LC_{50} 94.43 ppm, LC_{90} 344.93 ppm), and *Ocimum sanctum* leaf (LC_{50} 81.56 ppm, LC_{90} 541.80 ppm) (Kamaraj et al. 2008).

Chaithong et al. (2006) reported that the ethanol crude extracts of *Piper longum*, *Piper ribesoides*, and *Piper sarmentosum* showed strong larvicidal activity against larvae of *A. aegypti* with LC_{50} values of 2.23, 8.13, and 4.06 ppm and LC_{90} values of 4.80, 14.01, and 12.06 ppm, respectively. Similarly, the methanol leaf extracts of *Cassia fistula* (Mullai et al. 2008) and *Sida acuta* (Govindarajan 2010) showed larvicidal activity against larvae of *A. aegypti*. The efficacy of the ethyl acetate crude extracts of *Ervatamia coronaria* (Govindarajan et al. 2011), *Andrographis paniculata* (Perumalsamy et al. 2009), *Eclipta alba* (Rahuman et al. 2008a), and *Acalypha alnifolia* (Kovendan et al. 2012) against larvae of *A. aegypti* were investigated and found to have LC_{50} values of 89.59, 118.67, 154.88, and 164.34 ppm and LC_{90} values of 166.04, 225.29, 288.61, and 435.07 ppm, respectively.

Table 14.1 Larvicidal activity of crude extracts of *C. odollam* against *A. aegypti* after 24 h of exposure

Part used	Concentration (g/l)	Mortality (%)[a] ±SD	LC_{50} g/l (ULC-LLC)	LC_{90} g/l (ULC-LLC)
Control	0	0 ± 0	> 1.0 g/l	> 1.0 g/l
Rind	0.05	0 ± 0		
	0.1	1.6 ± 2.2		
	0.25	2.4 ± 2.2		
	0.5	4.8 ± 1.8		
	1	12.8 ± 3.3		
	2	18.4 ± 3.6		
Bark	0.05	1.6 ± 2.2	> 1.0 g/l	> 1.0 g/l
	0.1	1.6 ± 2.2		
	0.25	2.4 ± 2.2		
	0.5	5.6 ± 2.2		
	1	8 ± 4		
	2	12.8 ± 3.4		
Leaf	0.05	0.8 ± 1.8	> 1.0 g/l	> 1.0 g/l
	0.1	3.2 ± 3.3		
	0.25	3.2 ± 1.8		
	0.5	4 ± 2.8		
	1	8.8 ± 3.3		
	2	18.4 ± 2.2		
Seed kernel	0.05	1.6 ± 2.2	0.76 (0.48–1.1)	> 1.0 g/l
	0.1	3.2 ± 1.8		
	0.25	7.2 ± 3.3		
	0.5	25.6 ± 8.3		
	1	67.2 ± 5.2		
	2	72 ± 5.4		

Control—nil mortality. Significant at $p < 0.05$ level
[a]Mean value of five replicates

LC_{50} lethal concentration that kills 50% of the exposed larvae, LC_{90} lethal concentration that kills 90% of the exposed larvae, *UCL* upper confidence limit, *LCL* lower confidence limit

14.3.2 Larvicidal Activity of Fraction Derived from Seed Kernel and Subfractions Derived from Ethyl Acetate Fraction

Figure 14.2 shows the larvicidal activity of the ethyl acetate, n-hexane, and insoluble fractions derived from seed kernels. The larvicidal activity showed high larval mortality in the ethyl acetate fraction, a moderate effect in the n-hexane

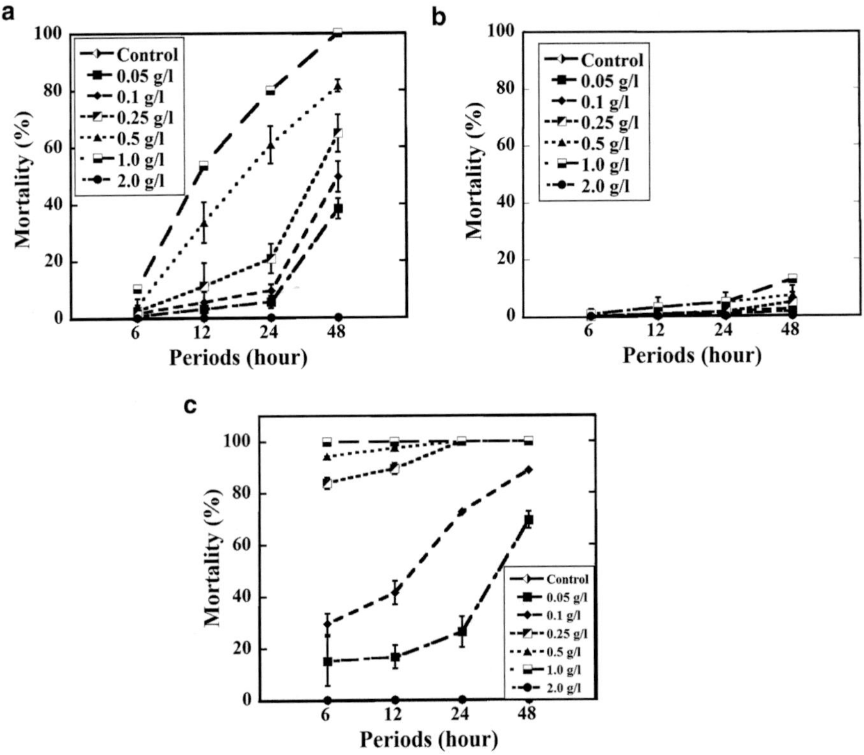

Fig. 14.2 Mortality of *A. aegypti* after exposure to fractions derived from seed kernels of *C. odollam*. (**a**) = n-hexane fraction, (**b**) = insoluble fraction, (**c**) = ethyl acetate fraction. Twenty-five larvae of *A. aegypti* were used per replication, and the data were averaged ($N = 5$). Error bars represent standard deviations

fraction, and a low effect in the insoluble fraction. The LC_{50} and LC_{90} values of the ethyl acetate fraction were 0.07 and 0.14 g/l, respectively, and these values for the n-hexane fraction were 0.42 and more than 1.0 g/l (Table 14.2). It is likely that some active compounds are concentrated in the ethyl acetate fraction. The larvicidal activity of the subfractions derived from the ethyl acetate fraction is shown in Fig. 14.3. Ten subfractions were obtained from the column chromatography stage. Subfractions 1 and 7 showed the highest larval mortality, namely, 100% larval mortality after 24 h of exposure (Fig. 14.3b). The bioactive compounds were probably present in subfractions 1 and 7. Furthermore, the isolation and purification of the bioactive compounds will be indispensable in future studies. Muthu et al. (Muthu et al. 2012) isolated the bioactive compound pectolinaringenin from *Clerodendrum phlomidis*, and it was assayed against the early fourth-instar larvae of the filarial vector *Culex quinquefasciatus* and the dengue vector *A. aegypti*. The results showed that the compound pectolinaringenin was effective against *Culex quinquefasciatus* (LC_{50} 0.62 ppm, LC_{90} 2.87 ppm) and *A. aegypti* (LC_{50} 0.75 ppm, LC_{90} 5.31 ppm).

Table 14.2 Larvicidal activity of fractions derived from seed kernels of *C. odollam* against *A. aegypti* after 24 h of exposure

Fractions	Concentration (g/l)	% mortality [a] ± SD	LC_{50} g/l (ULC-LLC)	LC_{90} g/l (ULC-LLC)	X^2
Ethyl acetate	Control	0.0 ± 0.0	0.07 (0.06–0.08)	0.14 (0.12–0.16)	1.4
	0.05	26.4 ± 10			
	0.1	72.8 ± 5.9			
	0.25	100 ± 0.0			
	0.5	100 ± 0.0			
	1.0	100 ± 0.0			
n-hexane	Control	0.0 ± 0.0	0.42(0.27–0.77)	> 1.0 g/l	12.3
	0.05	5.6 ± 2.2			
	0.1	9.6 ± 2.2			
	0.25	20.8 ± 5.2			
	0.5	60.8 ± 6.6			
	1.0	80 ± 8.5			

Control—nil mortality. Significant at $p < 0.05$ level
[a]Mean value of five replicates
LC_{50} lethal concentration that kills 50% of the exposed larvae, LC_{90} lethal concentration that kills 90% of the exposed larvae, *UCL* upper confidence limit, *LCL* lower confidence limit, $\chi 2$ chi-square

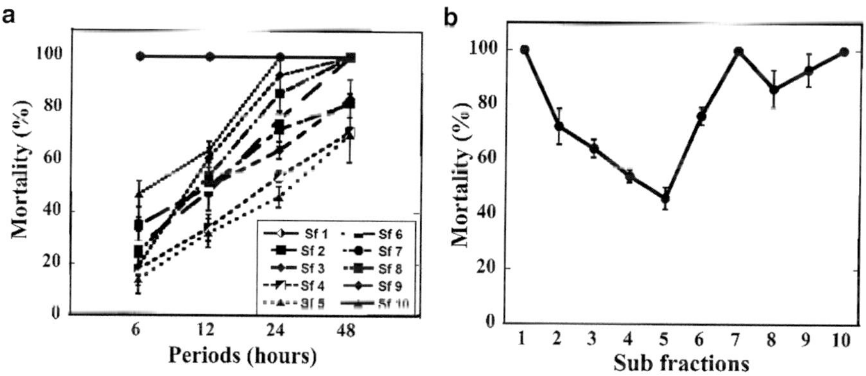

Fig. 14.3 Mortality of *A. aegypti* after exposure to subfractions derived from the ethyl acetate fraction of *C. odollam*. (**a**) = 48 h of exposure, (**b**) 24 h of exposure. Twenty-five larvae of *A. aegypti* were used per replication, and the data were averaged ($N = 5$). Error bars represent standard deviations

It was reported that oleic and linoleic acids isolated from *Citrullus colocynthis* showed larvicidal activity against *A. aegypti* (with LC_{50} values of 8.80 and 18.20, respectively, and LC_{90} values of 35.39 and 96.33 ppm, respectively), *Anopheles stephensi* (with LC_{50} values of 9.79 and 11.49 and LC_{90} values of 37.42 and 47.35 ppm, respectively), and *Culex quinquefasciatus* (LC_{50} values of 7.66 and 27.24 and LC_{90} values of 30.71 and 70.38 ppm, respectively) (Rahuman et al.

2008b). Emodin and pirimiphos-methyl isolated from *Cassia obtusifolia* seeds were strong against *A. aegypti* (LC_{50} 2.31 ppm), *A. togoi* (LC_{50} 2.67 ppm), and *Culex pipiens pallens* (LC_{50} 1.95 ppm) (Yang et al. 2003). Tectoquinone isolated from *Cryptomeria japonica* exhibited potent mosquito larvicidal activity against *A. aegypti* (LC_{50} 3.3 ppm, LC_{90} 8.8 ppm) and *A. albopictus* (LC_{50} 5.4 ppm, LC_{90} 26.9 ppm) (Cheng et al. 2008). Compounds of two new triterpenoids, 22, -23-dihydronimocinol and desfurano-6alpha-hydroxyazadiradione, were isolated from leaves of *Azadirachta indica* along with a known meliacin: 7alpha-senecioyl-(7-deacetyl)-23-O-methylnimocinolide showed mortality against the mosquito larvae of *Anopheles stephensi* with LC_{50} 60 and 43 ppm, respectively (Siddiqui et al. 2003).

Rahuman et al. (2008c) reported the bioactive compound of β-sitosterol isolated from *Abutilon indicum* as a potential new mosquito larvicidal agent against *A. aegypti* (LC_{50} 11.49 ppm, LC_{90} 47.35 ppm), *Anopheles stephensi* (LC_{50} 3.58 ppm, LC_{90} 15.03 ppm), and *Culex quinquefasciatus* (LC_{50} 26.67 ppm, LC_{90} 102.92 ppm). Rajkumar and Jebanesan (2010) also reported that the constituents of sabinene, biofloratriene, borneol, and β-bisabolol isolated from *Clausena dentata* exhibited significant larvicidal activity against *A. aegypti* (the LC_{50} values were 27.3, 47.4, 43.5, and 33.2 ppm, respectively). The bioactive component of β-thujaplicin isolated from *Chamaecyparis obtusa* leaves possessed high larvicidal activity against *A. aegypti* (LC_{50} 2.91 ppm), *Ochlerotatus togoi* (LC_{50} 2.60 ppm), and *Culex pipiens pallens* (LC_{50} 1.33 ppm) (Jang et al. 2005). The purified compound gymnemagenol isolated from the leaves of *Gymnema sylvestre* showed strong larvicidal activity against fourth-instar larvae of *A. subpictus* (LC_{50} 22.99 ppm) and *Culex quinquefasciatus* (LC_{50} 15.92 ppm) (Khanna et al. 2011). Pipernonaline derived from *Piper longum* also showed strong larvicidal activity against mosquito larvae of *Culex pipiens pallens* (LC_{50} 0.21 ppm, LC_{90} 0.52 ppm) (Lee 2000).

Earlier authors reported that the compound of gluanol acetate and a tetracyclic triterpene derived from *Ficus racemosa* exhibited LC_{50} and LC_{90} values of 14.55 and 64.99 ppm for *A. aegypti*, 28.50 and 106.50 ppm for *A. stephens*, and 41.42 ppm and 192.77 ppm for *Culex quinquefasciatus* (Rahuman et al. 2008d); vilasininoid and two havanensinoids isolated from *Turraea wakefieldii* and *Turraea floribunda* showed LD_{50} values of 7.1, 4.0 and 3.6 ppm, respectively, against larvae of *A. gambiae* (Ndung'u et al. 2004); 4-gingerol, 6-dehydrogingerdione, and 6-dihydrogingerdione isolated from *Z. officinale* showed LC_{50} values of 4.25 ppm, 9.80 ppm, and 18.20 ppm and LC_{90} values of 13.14 ppm, 37.42 ppm, and 96.33 ppm against larvae of *A. aegypti*, respectively, and showed LC_{50} values of 5.52 ppm, 7.66 ppm, and 27.24 ppm and LC_{90} values of 25.68 ppm, 30.71 ppm, and 70.38 ppm, respectively, against larvae of *Culex quinquefasciatus* (Rahuman et al. 2008e); compounds of cordiaquinones A and B, two novel meroterpenoid naphthoquinones named cordiaquinones J and K which were isolated from the roots of *Cordia curassavica*, were toxic against larvae of *A. aegypti* (Ioset et al. 2000).

In conclusion, a study to investigate the larvicidal activity of *C. odollam* extracts against a dengue vector, *A. aegypti*, has been carried out. Our study shows that *C.*

odollam extracts show potential for use in the control of *A. aegypti*. Identification of the components present in the active extracts is needed in future studies.

Acknowledgments The authors are grateful to Prof. Tsuyoshi Yoshimura and Dr. Khoirul Himmi Setiawan (RISH, Kyoto University) for their valuable advices.

References

Ahmed F, Amin R, Shahid I, Sobhani M (2008) Antibacterial, cytotoxic and neuropharmacological activities of *Cerbera odollam* seeds. Orient Pharm Exp Med 8:323–328

Bagavan A, Rahuman AA, Kamaraj C, Geetha K (2008) Larvicidal activity of saponin from *Achyranthes aspera* against *Aedes aegypti* and *Culex quinquefasciatus* (Diptera: Culicidae). Parasitol Res 103:223–229

Brown A (1986) Insecticide resistance in mosquitoes: a pragmatic review. J Am Mosq Control Assoc 2:123–140

Chaithong U, Choochote W, Kamsuk K, Jitpakdi A, Tippawangkosol P, Chaiyasit D, Champakaew D, Tuetun B, Pitasawat B (2006) Larvicidal effect of pepper plants on *Aedes aegypti* (L.) (Diptera: Culicidae). J Vector Ecol 31:138–144

Chapagain BP, Saharan V, Wiesman Z (2008) Larvicidal activity of saponins from *Balanites aegyptiaca* Callus against *Aedes aegypti* Mosquito. Bioresour Technol 99:1165–1168

Cheng S-S, Chang H-T, Chang S-T, Tsai K-H, Chen W-J (2003) Bioactivity of selected plant essential oils against the yellow fever mosquito *Aedes aegypti* larvae. Bioresour Technol 89:99–102

Cheng S-S, Huang C-G, Chen W-J, Kuo Y-H, Chang S-T (2008) Larvicidal activity of tectoquinone isolated from red heartwood-type *Cryptomeria japonica* against two mosquito species. Bioresour Technol 99:3617–3622

Choochote W, Tuetun B, Kanjanapothi D, Rattanachanpichai E, Chaithong U, Chaiwong P, Jitpakdi A, Tippawangkosol P, Riyong D, Pitasawat B (2004) Potential of crude seed extract of celery, Apium graveolens L., against the mosquito Aedes aegypti (L.) (Diptera: Culicidae). J Vector Ecol 29:340–346

Ciccia G, Coussio J, Mongelli E (2000) Insecticidal activity against *Aedes aegypti* larvae of some medicinal south American plants. J Ethnopharmacol 72:185–189

Coria C, Almiron W, Valladares G, Carpinella C, Ludueña F, Defago M, Palacios S (2008) Larvicide and oviposition deterrent effects of fruit and leaf extracts from *Melia azedarach* L. on *Aedes aegypti* (L.) (Diptera: Culicidae). Bioresour Technol 99:3066–3070

de Mendonça FA, Da Silva K, Dos Santos K, Júnior KR, Sant'Ana A (2005) Activities of some Brazilian plants against larvae of the mosquito *Aedes aegypti*. Fitoterapia 76:629–636

De Omena M, Navarro D, De Paula J, Luna J, De Lima MF, Sant'Ana A (2007) Larvicidal activities against *Aedes aegypti* of some Brazilian medicinal plants. Bioresour Technol 98:2549–2556

Dharmagadda V, Naik S, Mittal P, Vasudevan P (2005) Larvicidal activity of *Tagetes patula* essential oil against three mosquito species. Bioresour Technol 96:1235–1240

Gaillard Y, Krishnamoorthy A, Bevalot F (2004) *Cerbera odollam*: a 'suicide tree' and cause of death in the state of Kerala, India. J Ethnopharmacol 95:123–126

Govindarajan M (2009) Bioefficacy of *Cassia fistula* Linn.(Leguminosae) leaf extract against chikungunya vector, *Aedes aegypti* (Diptera: Culicidae). Eur Rev Med Pharmacol Sci 13:99–103

Govindarajan M (2010) Larvicidal and repellent activities of *Sida acuta* Burm. F.(Family: Malvaceae) against three important vector mosquitoes. Asian Pac J Trop Med 3:691–695

Govindarajan M (2011) Evaluation of *Andrographis paniculata* Burm. F.(family: Acanthaceae) extracts against *Culex quinquefasciatus* (say.) and *Aedes aegypti* (Linn.)(Diptera: Culicidae). Asian Pac J Trop Med 4:176–181

Govindarajan M, Karuppannan P (2011) Mosquito larvicidal and ovicidal properties of *Eclipta alba* (L.) Hassk (Asteraceae) against chikungunya vector, *Aedes aegypti* (Linn.)(Diptera: Culicidae). Asian Pac J Trop Med 4:24–28

Govindarajan M, Mathivanan T, Elumalai K, Krishnappa K, Anandan A (2011) Mosquito larvicidal, ovicidal, and repellent properties of botanical extracts against *Anopheles stephensi*, *Aedes aegypti*, and *Culex quinquefasciatus* (Diptera: Culicidae). Parasitol Res 109:353–367

Gubler DJ, Clark GG (1995) Dengue/dengue hemorrhagic fever: the emergence of a global health problem. Emerg Infect Diseases 1:55–57

Harve G, Kamath V (2004) Larvicidal activity of plant extracts used alone and in combination with known synthetic larvicidal agents against *Aedes aegypti*. Indian J Exp Biol 42:1216–1219

Hashim R, Boon J, Sulaiman O, Kawamura F, Lee C (2009) Evaluation of the decay resistance properties of *Cerbera odollam* extracts and their influence on properties of particleboard. Int Biodeter Biodegrad 63:1013–1017

Ioset J-R, Marston A, Gupta MP, Hostettmann K (2000) Antifungal and larvicidal cordiaquinones from the roots of *Cordia curassavica*. Phytochemistry 53:613–617

Jang Y-S, Jeon J-H, Lee H-S (2005) Mosquito larvicidal activity of active constituent derived from *Chamaecyparis obtusa* leaves against 3 mosquito species. J Am Mosq Control Assoc 21:400–403

Kamaraj C, Rahuman AA, Bagavan A (2008) Antifeedant and larvicidal effects of plant extracts against *Spodoptera litura* (F.), *Aedes aegypti* L. and *Culex quinquefasciatus* say. Parasitol Res 103:325–331

Kaushik R, Saini P (2008) Larvicidal activity of leaf extract of *Millingtonia hortensis* (family: Bignoniaceae) against *Anopheles stephensi*, *Culex quinquefasciatus* and *Aedes aegypti*. J Vector Borne Dis 45(1):66–69

Khanna VG, Kannabiran K, Rajakumar G, Rahuman AA, Santhoshkumar T (2011) Biolarvicidal compound gymnemagenol isolated from leaf extract of miracle fruit plant, *Gymnema sylvestre* (Retz) Schult against malaria and filariasis vectors. Parasitol Res 109:1373

Kiran SR, Bhavani K, Devi PS, Rao BR, Reddy KJ (2006) Composition and larvicidal activity of leaves and stem essential oils of *Chloroxylon swietenia* DC against *Aedes aegypti* and *Anopheles stephensi*. Bioresour Technol 97:2481–2484

Kovendan K, Murugan K, Vincent S (2012) Evaluation of larvicidal activity of *Acalypha alnifolia* Klein ex Willd.(Euphorbiaceae) leaf extract against the malarial vector, *Anopheles stephensi*, dengue vector, *Aedes aegypti* and Bancroftian filariasis vector, *Culex quinquefasciatus* (Diptera: Culicidae). Parasitol Res 110:571–581

Kumar MS, Maneemegalai S (2008) Evaluation of larvicidal effect of *Lantana camara* Linn against mosquito species *Aedes aegypti* and *Culex quinquefasciatus*. Adv Biol Res 2:39–43

Laphookhieo S, Cheenpracha S, Karalai C, Chantrapromma S, Ponglimanont C, Chantrapromma K (2004) Cytotoxic cardenolide glycoside from the seeds of *Cerbera odollam*. Phytochemistry 65:507–510

Lee SE (2000) Mosquito larvicidal activity of pipernonaline, a piperidine alkaloid derived from long pepper, *Piper longum*. J Am Mosq Control Assoc 16:245–247

Maheswaran R, Sathish S, Ignacimuthu S (2008) Larvicidal activity of *Leucas aspera* (Willd.) against the larvae of *Culex quinquefasciatus* say. And *Aedes aegypti* L. Int J Integr Biol 2:214–217

Markouk M, Bekkouche K, Larhsini M, Bousaid M, Lazrek H, Jana M (2000) Evaluation of some Moroccan medicinal plant extracts for larvicidal activity. J Ethnopharmacol 73:293–297

Morais SM, Cavalcanti ES, Bertini LM, Oliveira CLL, Rodrigues JRB, Cardoso JHL (2006) Larvicidal activity of essential oils from Brazilian croton species against *Aedes aegypti* L. J Am Mosq Control Assoc 22:161–164

Mullai K, Jebanesan A, Pushpanathan T (2008) Effect of bioactive fractions of *Citrullus vulgaris* Schrad. Leaf extract against *Anopheles stephensi* and *Aedes aegypti*. Parasitol Res 102:951

Muthu C, Reegan AD, Kingsley S, Ignacimuthu S (2012) Larvicidal activity of pectolinaringenin from *Clerodendrum phlomidis* L. against *Culex quinquefasciatus* say and *Aedes aegypti* L. (Diptera: Culicidae). Parasitol Res 111:1059–1065

Ndung'u MW, Kaoneka B, Hassanali A, Lwande W, Hooper AM, Tayman F, Zerbe O, Torto B (2004) New mosquito larvicidal tetranortriterpenoids from *Turraea wakefieldii* and *Turraea floribunda*. J Agric Food Chem 52:5027–5031

Park I-K, Lee S-G, Shin S-C, Park J-D, Ahn Y-J (2002) Larvicidal activity of isobutylamides identified in *Piper nigrum* fruits against three mosquito species. J Agric Food Chem 50:1866–1870

Pelah D, Abramovich Z, Markus A, Wiesman Z (2002) The use of commercial saponin from *Quillaja saponaria* bark as a natural larvicidal agent against *Aedes aegypti* and *Culex pipiens*. J Ethnopharmacol 81:407–409

Perumalsamy H, Kim N-J, Ahn Y-J (2009) Larvicidal activity of compounds isolated from *Asarum heterotropoides* against *Culex pipiens* Pallens, *Aedes aegypti*, and *Ochlerotatus togoi* (Diptera: Culicidae). J Medical Entomol 46:1420–1423

Prajapati V, Tripathi A, Aggarwal K, Khanuja S (2005) Insecticidal, repellent and oviposition-deterrent activity of selected essential oils against *Anopheles stephensi*, *Aedes aegypti* and *Culex quinquefasciatus*. Bioresour Technol 96:1749–1757

Promsiri S, Naksathit A, Kruatrachue M, Thavara U (2006) Evaluations of larvicidal activity of medicinal plant extracts to *Aedes aegypti* (Diptera: Culicidae) and other effects on a non target fish. Insect Sci 13:179–188

Rahuman AA, Gopalakrishnan G, Venkatesan P, Geetha K (2008a) Larvicidal activity of some Euphorbiaceae plant extracts against *Aedes aegypti* and *Culex quinquefasciatus* (Diptera: Culicidae). Parasitol Res 102:867–873

Rahuman AA, Venkatesan P, Gopalakrishnan G (2008b) Mosquito larvicidal activity of oleic and linoleic acids isolated from *Citrullus colocynthis* (Linn.) Schrad. Parasitol Res 103:1383–1390

Rahuman AA, Gopalakrishnan G, Venkatesan P, Geetha K (2008c) Isolation and identification of mosquito larvicidal compound from *Abutilon indicum* (Linn.) sweet. Parasitol Res 102:981–988

Rahuman AA, Venkatesan P, Geetha K, Gopalakrishnan G, Bagavan A, Kamaraj C (2008d) Mosquito larvicidal activity of gluanol acetate, a tetracyclic triterpenes derived from *Ficus racemosa* Linn. Parasitol Res 103:333

Rahuman AA, Gopalakrishnan G, Venkatesan P, Geetha K, Bagavan A (2008e) Mosquito larvicidal activity of isolated compounds from the rhizome of *Zingiber officinale*. Phytother Res 22:1035–1039

Rajkumar S, Jebanesan A (2010) Chemical composition and larvicidal activity of leaf essential oil from *Clausena dentata* (Willd) M. Roam.(Rutaceae) against the chikungunya vector, *Aedes aegypti* Linn.(Diptera: Culicidae). J Asia Pac Entomol 13:107–109

Sá RA, de Lima Santos ND, da Silva CSB, Napoleão TH, Gomes FS, Cavada BS, Coelho LCBB, Navarro DMdAF, Bieber LW, Paiva PMG (2009) Larvicidal activity of lectins from *Myracrodruon urundeuva* on *Aedes aegypti*. Comp Biochem Physiol C Toxicol Pharmacol 149:300–306

Siddiqui BS, Afshan F, Gulzar T, Sultana R, Naqvi SN-H, Tariq RM (2003) Tetracyclic triterpenoids from the leaves of *Azadirachta indica* and their insecticidal activities. Chem Pharm Bull 51:415–417

Tarmadi D, Prianto A, Guswenrivo I, Kartika T, Yusuf S (2007) The efficacy of *Carbera odollam* Gaertn and *Brugmansia candida* Pers against subterranean termite *Coptotermes* sp. J Wood Sci Tech 5:38–42

Utami S (2010) Laboratory evaluation on insecticidal activity of Bintaro (*Cerbera odollam* Gaertn) against Eurema spp. (in Indonesian). J For Plant Studies 7:211–220

Wandscheer CB, Duque JE, da Silva MA, Fukuyama Y, Wohlke JL, Adelmann J, Fontana JD (2004) Larvicidal action of ethanolic extracts from fruit endocarps of *Melia azedarach* and *Azadirachta indica* against the dengue mosquito *Aedes aegypti*. Toxicon 44:829–835

World Health Organization (2005) Guidelines for laboratory and field testing of mosquito larvicides. Communicable disease control, Prevention and eradication, WHO pesticide evaluation scheme. WHO, Geneva WHO/CDS/WHOPES/GCDPP/200513

Yang Y-C, Lee S-G, Lee H-K, Kim M-K, Lee S-H, Lee H-S (2002) A piperidine amide extracted from *Piper longum* L. fruit shows activity against *Aedes aegypti* Mosquito larvae. J Agric Food Chem 50:3765–3767

Yang Y-C, Lim M-Y, Lee H-S (2003) Emodin isolated from *Cassia obtusifolia* (Leguminosae) seed shows larvicidal activity against three mosquito species. J Agric Food Chem 51:7629–7631

Chapter 15
DNA-Based Occupancy Modeling: Human Activity Did Not Significantly Influence common palm civet's Occupancy

Sena A. Subrata, T.R.A. Budiman, and Fauzi Hamdan

Abstract The common palm civet (*Paradoxurus hermaphroditus*) may support natural forest restoration. They are long-distance dispersers of large seeds. Previous studies subtly indicated the influence of human activities on the occurrence of the civet. Here we reported the influence of human activities simultaneously with the availability of vegetative cover on the civet occurrence. We employed an occupancy modeling framework with covariates. We used DNA-based species identification to justify occupancy of a site. DNA was isolated from feces collected during fieldwork in a 600 ha secondary forest in Java during May–June 2015. We did not find any significant contribution of human activities on civet's occurrence. This finding supports previous study reporting that the civet is a disturbance tolerance species.

15.1 Introduction

Discontinuity of provisioning, regulating, and cultural services of ecosystems due to ecosystem degradation has been considered as a substantial threat to human population (Millennium Ecosystem Assessment 2005). Not only well-being but also our health, livelihood, and survival fundamentally depend on those ecosystem services (Costanza et al. 2014). The importance of those services cannot be neglected for the human population's persistence on the planet Earth.

Sustainability of ecosystem services is associated with plant diversity (Isbell et al. 2011); thus, ecosystems with high plant diversity, like tropical forests, are most likely capable of maintaining the ecosystem services. Therefore, continuing forest degradation is raising concern on unsustainability of ecosystem services and has driven forest restoration worldwide (Stanturf et al. 2014). The restoration has been commonly proposed to reverse the degradation and prevent further loss of the ecosystem services.

S.A. Subrata (✉) • T.R.A. Budiman • F. Hamdan
Faculty of Forestry, Gadjah Mada University, Jl. Agro No. 1, Yogyakarta 55281, Indonesia
e-mail: adisubrata@ugm.ac.id

B. McLellan (ed.), *Sustainable Future for Human Security*,
https://doi.org/10.1007/978-981-10-5430-3_15

Effective forest restoration encourages natural regeneration (Lamb et al. 2005). In the tropical forest, the regeneration of tree species depended in large part on seed dispersal carried out by animal (Corlett 1998). Particular animal species hold important ecological function as seed disperser of large-fruited tree species (Blackham et al. 2014). From a long list of animal species described as seed disperser in tropical forest (Corlett 2002), civet species was frequently reported as effective seed disperser of large-fruited tree species (Subrata and Syahbudin 2016; Lindsell et al. 2015; Nakashima and Sukor 2010). One of the civet species that is considered to play an important role in forest regeneration is the common palm civet (*Paradoxurus hermaphroditus*).

This species swallows seeds and defecates them in intact form. The seeds retain a good germination power after being defecated by the civet (Nakashima et al. 2010a, b). This animal species was reported dispersing large seeds across long distances, frequently between forest patches (Nakashima and Sukor 2010). They are considered to have an important role in plant colonization and ecosystem restoration. Despite its important role, only a few studies have reported the ecology of this species, particularly on their behavior. Their nocturnal trait was most probably responsible for the limited studies. Innovative techniques have been needed, at least to reliably detect their occurrence, in order to understand behavioral ecology of this species.

As with other wildlife species, the occurrence and behavior of the civet are influenced by the availability of food, shelter, predators, and most likely human activities. As this species is a food generalist, food availability is rarely considered as a limiting factor of their occupancy. Instead, shelter is an issue for their occupancy. The species prefers trees with a dense canopy for their resting site, although they are also occasionally observed resting on the ground where there is a dense vegetation mat (Nakashima et al. 2013). Although the civet was considered to be tolerant to disturbance (Nakashima et al. 2010b), however, their behavior may shift due to their high adaptability. They changed their behavior as a response to predation pressure including human presence (Joshi et al. 1995). As the civet may perceive human activities as disturbance or predation, it may influence civet's site occupancy. All of this evidence may indicate that cover is an important resource for the species occupancy. Nevertheless, there is still unclear information on which specific cover is preferred by this species in relation with human presence as a disturbance.

This information is important since forest restoration projects are frequently located near human activities. The project will receive much benefit from the civet if the species would like to occupy restored forest. Considering the benefit, we attempt to reveal factors influencing civet occupancy focusing on cover availability, human disturbance, and habitat type. We conducted this study to evaluate influence of human activities and vegetative cover simultaneously on civet's site occupancy.

15.2 Methods

We selected a secondary forest near human settlement in Petungkriyono, Central Java, Indonesia, as a research site. Local people grow coffee and other agricultural commodities beneath the forest canopy in some parts of the forest (Harimurti 2015). Meanwhile, previous studies reported the occurrence of the common palm civet in the surrounding areas (Rode-Margono et al. 2014). It is likely that there was interaction between local people and the civet in our research site.

This study was conducted within a framework of single-session occupancy modeling with covariates. Site occupancy was evidenced by civet's feces found at the site. Vegetation conditions and human presence were the habitat attributes or covariates. Fieldwork covering 600 ha of the forest was conducted in May–June 2015 to collect civet's feces and measure habitat attributes. Sixty grids (10 ha each) were distributed systematically in the area (Fig. 15.1). Within each grid, five to ten transects of 100 m length were established to search for and collect civet's feces. We treated each transect as a sample to justify probability of occupancy (POA) of civet (MacKenzie et al. 2006).

When found, feces were preserved in ethanol of concentration 95% until DNA isolation. DNA was isolated from civet feces using QIAamp mini stool kit (Qiagen) with a minor modification. The modification involved DNA incubation in ASL buffer before DNA isolation. Isolated DNA was amplified using markers designed specific for this species and KAPA2G Fast PCR Kits (Kappa) (Subrata 2015) and visualized on agarose gel. We used bands showed on the gel as evidence to justify occupancy of transect by civet.

Besides collecting feces, we also assessed habitat type, indications of human presence, and vegetative cover during the fieldwork. We classified habitat type into three categories: habitat with wild coffee plants (no indication of human intervention on the plants), habitat with coffee plants grown by farmers, and habitat with no coffee plants. We considered coffee plants as a clue for habitat categorization because the civet feeds on coffee cherry frequently when available. We detected the presence of human activities by observing the occurrence of tracks, huts, and residues. Vegetative cover was measured as horizontal coverage of vegetation at 30–100 cm above ground using a density board (Noon 1981).

We posed the latter three variables as covariates in an occupancy modeling using PRESENCE v. 2.4 (Hines 2006). We did not consider fruit availability as a factor affecting civet occurrence as fruit is highly abundant in the forest and the civet is a generalist species. The modeling was employed to reveal the influence of habitat type, vegetative cover, and human presence on civet's occupancy.

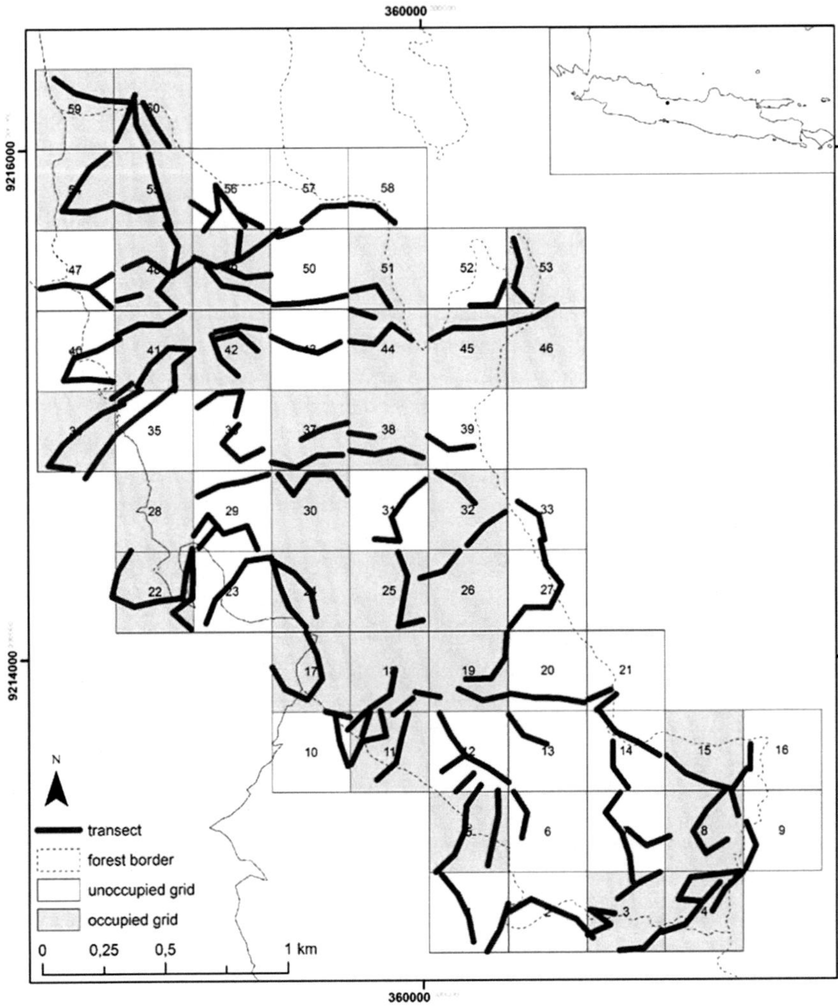

Fig. 15.1 Civet evenly occupied approximately half of a forest area in Petungkriyono, Central Java

15.3 Results and Discussion

We found 103 civet-like feces in 58 of 278 observed transects. We successfully identified 87 of them as civet feces using molecular analysis. Those feces were found in 32 of 60 grids. (Fig. 15.1). Based on the probability of occupancy (naive occupancy estimation = 0.53) and its spatial distribution, it appears that the civet is evenly distributed in the forest. This finding is similar with distribution pattern of the species in palm oil plantations in Sumatera (Jennings et al. 2015). In that study, Jennings et al. (2015) reported that common palm civet occurred extensively in oil

Table 15.1 Model selection: contribution of habitat type (hab), human activities (human), and vegetative cover (cover) in determining probability of civet's site occupancy

No	Model	AIC	ΔAIC	AIC weight	Model likelihood	Nr. Par.	2*Log likelihood
1	Ψ(cover),p(.)	285.60	0	0.4612	1	3	279.60
2	Ψ (human,cover),p(.)	287.44	1.84	0.1838	0.3985	4	279.44
3	Ψ (.),p(.)	287.56	1.96	0.1731	0.3753	2	283.56
4	Ψ (hab,cover),p(.)	289.19	3.59	0.0766	0.1661	5	279.19
5	Ψ (human),p(.)	289.55	3.95	0.064	0.1388	3	283.55
6	Ψ (hab),p(.)	291.14	5.54	0.0289	0.0627	4	283.14
7	Ψ (hab,human),p(.)	292.83	7.23	0.0124	0.0269	5	282.83

AIC Akaike information criterion, *Nr. Par.* number of parameters

palm plantations. It may indicate that the civet is highly tolerant to various habitat conditions and insensitive to human presence. This trait is favorable to support its ecological function for seed dispersal in the context of forest restoration. Natural forest regeneration is most likely assisted by this species.

Our further analysis showed that civet's occupancy was best modeled by vegetation cover only. Compared to other models involving human and/or habitat type, Akaike information criterion (AIC) of a model that involves only vegetative cover was the lowest (Table 15.1). The difference of AIC value between the first and the second model (ΔAIC) was quite large (1.84); therefore, we decided the first model as the best-suited model for explaining civet's occupancy. It means that among factors considered as habitat factors in the previous section, vegetation cover best explains why a site is occupied by the civet. Since we employed AIC value-based modeling, it did not mean that other factors or combinations of it have no substantial contribution to the civet's occupancy. This modeling opens possibility that the other factors may contribute, but vegetation cover explains best the occupancy. Our result may imply that forest restoration will need undergrowth vegetation below 1 m in order to provide shelter for its seed-dispersal agent.

Involving human activity into the model did not improve goodness of fit of the model. Our study showed that involvement resulted in second-best model (Table 15.1). This finding clarified the previous study that subtly indicated tolerance trait of this species. Several studies reported that the species survives well in human-modified areas (Nakashima et al. 2010b; Bartels 1964); meanwhile, other studies indicated that the tolerance trait of this species is closely associated with their nocturnal habits (Lindsell et al. 2015; Nakashima et al. 2010a) without any further information on how human affects their population or behavior. Our study underlined that human presence unlikely influences the civet occurrence. In line with Jennings et al. (2015) and Nakabayashi et al. (2014), we argue that temporal niche segregation between human and the civet is the reason behind a weak contribution of human presence on the civet occupancy model. In our study site, local people were assumed to actively care their agricultural plant during the day. Since the civets may perceive human as a source of predation, the civet may adapt to the day situation by avoiding human. So, we presumed their nocturnal trait is most likely associated with adaptation of day situation, considering their well-adapted behavior (Joshi et al. 1995). It warrants further research questioning whether physical condition of the habitat related to the solar radiation or predation presence determines activity of the civet. This information has importance in forest restoration to optimize temporal resource sharing between human and civet.

The vegetative cover has significant contribution in explaining civet's site occupancy (covariate coefficient $= -1.16$; SE $= 0.76$). Our data showed that average vegetation cover (horizontal) in our research site was moderate ($\bar{X} = 41.64\%$; SD $= 16.32$). It implies that a site with moderate density of vegetation cover has high probability to be occupied by the civet. The civet avoids very dense vegetation cover at 30–100 cm. This finding clarifies previous study reporting resting site of the civet in the forest canopy (Bartels 1964) and on the ground (Nakashima et al. 2013). In the later study, Nakashima et al. (2013) observed resting site of the species on the ground with dense vegetation mat for daybeds. However, there was no further information available on vegetation cover surrounding the site. We consider this coverage is important in determining occupancy. Although important, no prior studies have evaluated this factor in detail. Jennings et al. (2015) suggested that the availability of suitable resting site encourages habitat utilization of the civet. We view that preventing sight from predators is a key characteristic of most animal shelter, including civet. However, as the civet rests at daytime when humans are active, using undergrowth vegetation below 1 m as shelter is suboptimal. It is likely that the civet uses the undergrowth vegetation as a temporary shelter, not permanent. The civet may use it as a refugee or temporary shelter to ease their access to food sources. It is likely that the civet selects other sites as a permanent resting site, such as tree crown, as indicated by Bartels (1964). This implies that high trees surrounding forest restoration area are important, besides undergrowth vegetation, to facilitate natural regeneration assisted by the civet.

Methods and techniques applied in this study are able to provide reliable data on factors influencing civet's occupancy. Comparing to previous study, we incorporated imperfect detection of civet's occurrence in determining factors contributing to occupancy. We elaborated imperfect detection by spatially repeating observation of the civet occurrence and involving it into occupancy modeling. Additionally, we introduced molecular techniques to reliably justify civet occurrence in an inexpensive way. This DNA-based species identification resulted in reliable data without direct observation of the species. Thus, the field survey can be conducted in a convenient way that provides good quality of data. This identification technique saved much resource and avoided misidentification. Previous studies (Nakashima et al. 2013) have analyzed present-only data acquired using radiotelemetry to reveal resource selection of this species. It treated pseudo- and false absent as true absent that may lead to biased conclusion (Millennium Ecosystem Assessment 2005). Additionally, this radiotelemetry observation technique demands huge resources, including time, person, and fund, and is risky because nocturnal trait of the civet needs night observation. The technique is capable to collect spatial and temporal ecological data at very fine spatial and temporal resolution. But, it may be cost prohibitive to be applied in tropical situations, where most tropical developing countries have very limited resources for research.

In summary, although our study found that human presence was not a robust factor influencing the civet occupancy, however, the presence may come into effect when humans or the civet shift their habits so that temporal niche segregation does not exist anymore. Ecosystem modification in order to restore a forest should create suitable resting sites and/or temporal niche segregation between humans and the civet. Using these methods, the ecosystem may support restoration projects naturally, and sustainable resources may be expected.

References

Bartels E (1964) On Paradoxurus hermaphroditus javanicus (Horsfield, 1824): The Common Palm Civet or Tody Cat in Western Java – notes on its food and feeding habits and its ecological importance for wood and rural biotopes. Beaufortia 10:193–201

Blackham GV, Webb EL, Corlett RT (2014) Natural regeneration in a degraded tropical peatland, Central Kalimantan, Indonesia: implications for forest restoration. For Ecol Manag 324:8–15

Corlett RT (1998) Frugivory and seed dispersal by vertebrates in the Oriental (Indomalayan) Region. Biol Rev Camb Philos Soc 73:413–448

Corlett RT (2002) Frugivory and seed dispersal in degraded tropical east Asian landscapes. In: Levey DJ, Silva WR, Galetti M (eds) Seed dispersal and frugivory: ecology, evolution, and conservation. CABI Publishing, New York, pp 451–465

Costanza R, de Groot R, Sutton P et al (2014) Changes in the global value of ecosystem services. Glob Environ Change 26:152–158

Harimurti (2015) Statistik Daerah Petungkriyono tahun 2015. Badan Pusat Statistik Kabupaten Pekalongan, Pekalongan

Hines J. (2006) PRESENCE2-Software to estimate patch occupancy and related parameters.

Isbell F, Calcagno V, Hector A et al (2011) High plant diversity is needed to maintain ecosystem services. Nature 477:199–202

Jennings AP, Naim M, Advento AD et al (2015) Diversity and occupancy of small carnivores within oil palm plantations in central Sumatra, Indonesia. Mammal Res 60:181–188

Joshi AR, Smith JLD, Cuthbert FJ (1995) Influence of food distribution and predation pressure on spacing behavior in palm civets. J Mammal 76:1205–1212

Lamb D, Erskine PD, Parrotta JA (2005) Restoration of degraded tropical forest landscapes. Science 310:1628–1632

Lindsell JA, Lee DC, Powell VJ et al (2015) Availability of large seed-dispersers for restoration of degraded tropical forest. Trop Conserv Sci 8:17–27

MacKenzie DI, Nichols JD, Royle AJ et al (2006) Occupancy estimation and modeling: inferring patterns and dynamics of species occurrence. Academic, Amsterdam

Millennium Ecosystem Assessment (2005) Ecosystems and human well-being: synthesis. Island Press, Washington, DC

Nakabayashi M, Nakashima Y, Bernard H et al (2014) Utilisation of gravel roads and roadside forests by the common palm civet (*Paradoxurus hermaphroditus*) in Sabah, Malaysia. Raffles Bull Zool 62:379–388

Nakashima YN, Sukor JA (2010) Importance of common Palm civet as (*Paradoxurus hermaphroditus*) as a long-distance disperser for large-seeded plants in degraded forests. Tropics 18:221–229

Nakashima Y, Inoue E, Inoue-Murayama M et al (2010a) Functional uniqueness of a small carnivore as seed dispersal agents: a case study of the common palm civets in the Tabin Wildlife Reserve, Sabah, Malaysia. Oecologia 164:721–730

Nakashima Y, Inoue E, Inoue-Murayama M et al (2010b) High potential of a disturbance-tolerant frugivore, the Common Palm Civet *Paradoxurus hermaphroditus* (Viverridae), as a seed disperser for large-seeded plants. Mammal Study 35:209–215

Nakashima Y, Nakabayashi M, Sukor JA (2013) Space use, habitat selection, and day-beds of the common palm civet (*Paradoxurus hermaphroditus*) in human-modified habitats in Sabah, Borneo. J Mammal 94:1169–1178

Noon BR (1981) Techniques for sampling avian habitat. In: The use of multivariate statistics in studies of wildlife habitat. Rocky Mountain Forest and Range Experiment Station, Colorado, pp 42–52

Rode-Margono E, Voskamp A, Spaan D et al (2014) Records of small carnivores and of medium-sized nocturnal mammals on Java, Indonesia. Small Carniv Conserv 50:1–11

Stanturf JA, Palik BJ, Williams MI et al (2014) Forest restoration paradigms. J Sustain For 33:161–194

Subrata SA (2015) Using species specific primers for detecting DNA in a wildlife feces. In: The proceeding of international seminar on natural resources biotechnology: from local to global, Atmajaya University Yogyakarta

Subrata SA, Syahbudin A (2016) Common Palm Civet as a potential seed disperser of important plant species in Java. AIP Conf Proceeding, 4

Chapter 16
Sustainable Development of Traditional Processed Fishery Products: A Feasibility and Risk Analysis of Small-Scale Businesses in West Kalimantan

Eva Dolorosa and Nurliza Nurliza

Abstract This study was conducted to determine the profitability, risk, and financial feasibility of traditional processed fishery products for sustainable development in Sambas, West Kalimantan. The primary aim of the study is to determine the business opportunity and to integrate poor households into markets. Profit analysis and coefficient variance were used for profitability and risk analysis; meanwhile, net present value (NPV), internal rate of return (IRR), benefits to costs ratio (B/C), and payback period (PP) were used for financial feasibility analysis. The results of the study show that all the indicators for financial feasibility analysis are positive and accepted, but they have high to very high risks commonly associated with raw materials, production process, prices, and markets. This analysis shows the potential for adding values to the small-scale fishery businesses through processed fishery products. Thus, the government and private industry should find the best technologies, practices, regulation, and management. They should coordinate institutional development strategies and local initiatives to ensure that the growth is sustainable. Public interventions should provide adequate extensions and advisory services for supporting the governance and development of small-scale fishery processed products. Sharing experience, in this regard, can help to speed the development of this sector.

16.1 Introduction

Indonesia is an archipelagic country which has a lot of fishery products coming from small-scale and traditional fisheries strongly influenced by local knowledge and local practices. The fishery sector has a big potential for processed fishery products. The trend of Indonesian fishery production has grown in the last 5 years,

E. Dolorosa (✉) • N. Nurliza
Agribusiness Department, Faculty of Agriculture, Tanjungpura University, Jl. Prof. Dr.Hadari Nawawi No.1, Pontianak 78124, Kalimantan Barat, Indonesia
e-mail: edolorosa@gmail.com

B. McLellan (ed.), *Sustainable Future for Human Security*,
https://doi.org/10.1007/978-981-10-5430-3_16

amounted to 23.22% with an average value of Rp.145 trillion, equivalent to approximately (The Center for Data Statistic and Information 2014) US$ 11 billion. Fishery has an important role as a source of nutritious food, employment, and revenue for people in many countries. The production of fisheries is influenced by environmental conditions, technology, and economic climate. Most fish supply comes from marine capture fishery, which comprises some types of fishery such as industrial, small-scale, and traditional fisheries. Small-scale fishery businesses are important, but they are often undervalued. In the developing world and rural areas, nearly 90% of the full-time or part-time fishers are within the small-scale sectors (CFS 2014).

Fish is a major source of protein. The harvesting, handling, processing, and distribution of fish provide livelihood for millions of people and provide foreign exchange earning to many countries (Al-Jufaili and Opara 2006). While fishery production is often abundant, the market demand fluctuates widely. So, preservation and processing of fishery products are required in order to maintain the quality and nutrition of fish and to ensure that the fish is safe for human consumption. Appropriate processing of fish enables the maximal use of raw materials and production of value-added products which are obviously the basis of processing profitability (Tawari and Abowei 2011).

Apart from the primary production sector, fisheries provide numerous jobs in ancillary activities such as processing, packaging, marketing and distribution, manufacturing of fish processing equipment, net and gear making, boat construction and maintenance, research, and administration. All of this employment, together with the dependents, are estimated to support the livelihoods of hundreds of millions of people.

The Coherence for Development (CODE) report explores the role of fisheries and aquaculture for sustainable development, economic growth, and global food security. It also examines how policies related to trade, governance, and regulatory regimes impact on the fisheries sector and can maximize the benefits of the development in a sustainable manner (Martini 2013).

However, at this time, the level of utilization of the fishery catch, especially the noneconomic one, is still not optimal because the utilization is still limited to traditionally processed and fresh fish products. As a result, the fish are not handled well in the boat; so, the fish are landed with low quality (20–30%) which contributes to the high rate of loss (losses) reaching 30–40%. Due to the losses of fishery products, fisheries processed products for the family become food supply, provide added value and also increase the household income of fishermen. The processed fishery products are divided in two categories: modern and traditional processed fishery products. The focus of this study is the traditional fishery products (TFPs) which are usually produced by applying traditional preserving methods such as salting, fermenting, drying, and smoking. Although such processing methods have been known as traditional techniques for many years, they are still widely used around the world because of their specific taste and aroma (Köse 2010).

Sambas district is one of the coastal districts in West Kalimantan, Indonesia. It has a long beach line reaching 198.36 km^2 and the potential fishery around

23.25 tons/year. In order to overcome the abundance of fish production, local people had processed fisheries products that can be consumed and sold. However, the local communities have low knowledge and technology. They use a simple and easy processing such as processing fish into salted fish, dried fish, fish crackers, and *terasi* (shrimp paste). Thus, the objectives of this research are to identify the processing and marketing and to analyze the feasibility and the risk of this small-scale business because the aggregate of small-scale industries has the potential for accelerating the pace of economic development and employment generation of the country. Additionally, it has successfully played a positive role in the economy of the rural areas and has been sustainable for a long term.

16.2 Literature Review

16.2.1 The Importance to Employment and Poverty Reduction

Postharvest activities of fishing play a crucial role in a number of developing countries. They contribute to job creation and food security, both directly and indirectly. In a number of LDCs, fish provide more than half of the animal protein consumed, including Bangladesh, Cambodia, Gambia, Guinea, Ghana, Indonesia, Sierra Leone, and Togo, among others (Béné 2006). Of course, the incomes earned from selling fish and fishery processed are also important even though people engaged in artisanal fishing and fish processing tend to have low incomes (Béné 2006; Béné et al. 2010).The term "artisanal" refers to the relative level of technology, while "small scale" refers to the size of fishing unity (scale). According to FAO glossary (FAO 1999), artisanal or small-scale fisheries are traditional fisheries involving fishing household (as opposed to commercial companies), using relatively small amount of capital and energy, relatively small fishing vessels (if any), making short fishing strips, close to shore, and mainly for local consumption. They can be for subsistence or commercial.

As for the informal sector more generally, artisanal fishing is a survivalist activity. Artisanal fishing is a major source of employment and earnings, but it is handicapped by rudimentary infrastructure and poor hygiene. Foreign vessels using advanced technologies to catch high-value species dominate the fishing industry.

Industrial fishing operations provide revenues to LDC governments through fishing agreements, but they often land little of their catch in the LDCs and sometimes contribute to depletion of stocks. The importance of artisanal fishing further complicates the policy toward the fishing sector, as there may be trade-offs between employment and resource management (Golub and Varma 2014).

A further characteristic of LDCs' fishing industries is the coexistence of industrial and artisanal fishing. Artisanal fishing, like the informal economy more generally, provides employment of last resort. In fact, the common-pool resource aspect and low capital intensity of fishing enable easy entry of low-skilled people with few other options. They distinguish a labor-buffer effect of absorbing chronic surplus labor from the safety net effect of fishing on short-term shocks. Nevertheless, artisanal fishers are highly vulnerable due to (1) high exposure to risks, (2) high sensitivity to those risks, and (3) low capacity to adapt to risks. The risks include: physical risks (drowning, accidents), weather-related risks (tropical storms, tsunamis, floods), possibly exacerbated by climate change, and resource risks as fish stocks can migrate or be subject to overfishing or disease (Béné et al. 2010).

The wealth-based model to poverty alleviation for fisheries focuses on increased investment, value added, and exports (Béné 2006); meanwhile, the welfare model focuses on the safety net and labor-buffer effects on sustaining incomes of the poor, and there is no evidence that the creation of processing factories reduces employment of artisanal (Golub and Varma 2014).

Furthermore, the majority of jobs in fisheries are in postharvest distribution and processing, and women tend to dominate these activities, particularly when they are artisanal. Women also frequently provide funds to invest in the family fisheries business. Despite women's substantial and increasing involvement in fisheries in some countries, gender inequities arise from traditional beliefs and customs and present-day legal and regulatory barriers (Golub and Varma 2014).

16.2.2 The Importance of Small-Scale Enterprises

The maintenance of small-scale enterprises is a widespread policy objective in many countries. The contributions of small-scale enterprises to employment generation in Nigeria were measured by providing sectorial analysis of the efficacy of small-scale enterprises as a vibrant tool for employment generation (Kadiri 2012). In fact, small-scale fisheries contribute about half of global fish catches and employ over 90% of the world's capture fishers and fish workers; also considering fish processing and other related jobs, about half of the persons employed are women (FAO 2013).

In many marginalized coastal areas of the world, small-scale fisheries represent a key economic activity of direct and indirect employments and provide nutritious food for the local population. Furthermore they tend to be strongly anchored in local communities, often reflecting historic links to adjacent fishery resources, traditions, and values and supporting social cohesion. In contrast, in some regions, such as in EU, regional income and employment generated by small-scale fisheries are very low compared to the other economic sectors (Macfadyen et al. 2011). Despite the marginal impact on the economy and the general trends indicating that fishing activity (especially small-scale fisheries) is decreasing in EU, the recognition of the social and cultural role of small-scale fishermen was explicitly included in the European Commission's Green Paper on the reform of the common fisheries policy (CFP). On the other hand, this low level of investments is probably a characteristic allowing a higher adaptability and resilience of fishing communities. Because economic risks are reduced and there are fewer restraints in work mobility, low investments also involve difficulties in getting new technologies and consequently low productivity and low capacity to meet market requirements (Sacchi 2011).

Actually, knowledge of small-scale fisheries is still limited (Macfadyen et al. 2011). As indicated by the literature, small-scale fisheries generally attain higher prices than large-scale fisheries. This may be explained by both the differences in quality, linked to the freshness of the products and trip duration, by the marketing channels, and by the steps taken to better identify the products. The relatively low quantities landed by most small-scale fisheries also allow the crew to devote more time to cleaning and preparing the landings for favorable presentation and that is likely to be more richly rewarded products. Distinctions, however, should be made case by case. In fact, the small size of the vessels may inhibit onboard handling and storage facilities and that may have negative connotations, which might even reduce the quality of the product. In other cases, the absence of appropriate infrastructures in the fishing ports may have the same results, determining the first weak point of small-scale fisheries products in the value chain. Marketing channels for some live products are logistically difficult requiring significant investment in onboard but especially in onshore storage facilities. Finally, in some cases, the large volume of landings of large-scale fisheries on the markets may seasonally have a significant impact on the price of the products also landed by the small-scale fisheries. Price impact of illegal landings is not to be underestimated. In other cases (where product is addressed to export), price of products is subject only to international price drivers (IFREMER 2007).

The scarcity of resources, the low investments of small enterprises, the weak market position, and the sanitary and safety problems linked to new distribution markets are all elements causing important difficulties for small-scale fishermen (Malorgio et al. 2014). On the other hand, it is important to give specific emphasis to remote areas where alternative activities to fishery are not easily identifiable, and where small-scale fishing still represents the main or even the only work opportunity. In this case, it is possible to consider the relevance of both infrastructures for the integration of landings in the market (i.e., development of auctions, vertical and horizontal cooperation) and infrastructures for improving the quality of life of fishermen communities.

16.2.3 Challenges of Small-Scale Industries for Sustainability Development

The empirical research on the impact of financing small-scale enterprises on economic growth in Nigeria reveals that loan to small-scale enterprises has a positive impact on the economic performance while interest rate has a negative impact on economic growth. They concluded that the worst obstacles confronting small-scale industries in Nigeria are managerial capacity and access to capital or finance (Onakoya et al. 2012).

The problems of financing small-scale business enterprises are identified as the sources of finance, types of finance available for small business enterprises, and problem inhibiting business enterprises in Nigeria in securing funds for the smooth operation (Ighoroje and Oboro 2011). It concluded that adequate finance is indispensable for the successful operations of small-scale enterprises. The financial management skills which form the bedrock upon which the success of any business can rest and grow are highly needed. The other skills which are also highly needed are administrative skills, professional skills, and information technology skills (Ikeanyionwu 2006).

Scholars have indicated that starting a business is a risky venture and warn that the chances of small-scale business owners making it pass the 5-year mark are very small. Despite the efforts, support, and incentive programs for small-scale industries, it seems reasonable to expect that small-scale business would grow and flourish, but the rate of business failure continues to increase, creating unemployment problems and poverty (Akabueze 2002).

Sustainable development as an evolving concept, which is continually redefined and reinterpreted, can be formally stated as the twin principles of intra- and intergenerational equity (World Bank 2006). The World Council for Environment and Development (WSSD 2002) defines sustainable development as development that meets present needs without compromising the ability of future generation to meet their own needs. In other words, it is the development that can stand a test of time. Sustainability implies the need to depart from hither to dominant models of development that fail to balance the need of people and the planet in the pursuit of peace and prosperity (Wals 2009). The sustainability debate is anchored on the assumption that societies need to manage three types of capital (economic, social, and natural), which may be non-substitutable and whose consumption might be irreversible (Dylick and Hockerts 2002). It emphasizes the importance of capacity building that is an externally or internally initiated process designed to help individuals and groups appreciate and manage their changing circumstances, with the objective of improving the stock of human, social, financial, physical, and natural capital in an ethically defensible way.

The aggregate of small-scale industries has the potential for accelerating the pace of economic development and employment generation of the country, and has successfully played a positive role in the economy of rural areas. For small-scale industries to perform such functions effectively, these challenges in the business environment must be addressed. The obstacles identified and examined in the study include: credit not often assessable, lacking competence and managerial skills, information technology, and investment in research and development. Others are bad credit record; corruption, crime, and security of life and property; high cost of doing business; high competition; and weak infrastructural base.

Based on these findings, this research recommends that the financial system is urged to extend credit facilities with minimal interest rates as a means of resolving the challenges of credit not often assessable and empowering emerging small-scale industries through stimulus packages. Furthermore, there is a need for the various tiers of government to deal decisively with crime and corruption in both public and business communities, and to streamline the multiple taxation and improvement in infrastructural facilities (Onukwuli et al. 2014).

16.2.4 The Role of Innovation to Enhance Production and Market for Sustainability Development

The enhancement of production and market can be through individual or collective strategies. In many cases, individual strategies are infeasible. At the same time, pooling the supply is the only way of increasing clout in negotiation with buyers. This is necessary in the fishing industry, particularly for small-scale enterprises, where the production flow of each boat is generally low and discontinuous, and where the product is highly perishable and cannot be either standardized or differentiated. Aggregation and cooperation are then required to reach an adequate volume and control on supply, allowing cost reduction, supply chain synergies, product differentiation, and sales policy that will ensure a more profitable placement of local products. Here, we can consider the forms of horizontal and vertical cooperation and broader partnerships with public authorities, civic society, and enterprises.

Business cooperation is the collaboration between juridical and economically independent companies to raise the common competitiveness. As a general objective, such strategies can lead to lower costs and higher efficiency; customer expectations can be met more easily; synergies are generated in order to pool resources and to share strength and know-how; stability and sustainability of the market is gained. From a juridical and technical perspective, the forms of cooperation include: joint ventures, cooperatives, consortia, service agreements, etc. Each of these forms has its own typical characteristics with pros and cons, and should be considered carefully depending on business elements. Furthermore, these cooperation forms can be characterized by differences in the geographical extension, in terms of the duration and the field of cooperation (i.e., R&D, distribution, purchasing, marketing, production) (Malorgio et al. 2014).

In fact, on a broader perspective (FAO 2013), small-scale fishing communities need access to the full spectrum of financial, social, and institutional services and resources to sustain their livelihoods, and public organizations should support investment in human resource development such as health, education, literacy, and other skills training. The scale and priorities of this state intervention may change considerably depending on the location, but public institutions should ensure that small-scale fishing communities have access to essential public services, starting from decent housing, sanitation, potable water, and electricity. It should also be guaranteed that small-scale fishers and fish workers are covered by unemployment insurance and social security schemes with benefits equal to other professional groups in the country.

16.3 Methodology

Quantitative data for the study were collected through a cross-sectional survey involving 90 producers of traditional processed fishery products during April–May 2015 in six subdistricts in Sambas, West Kalimantan, that has the largest marine fishery potential. Traditional processed fishery products consist of salted fish, dried fish, fish crackers, and *terasi*. A questionnaire was developed and further pretested in a pilot study. It measured a wide variety of constructs including risk and feasibility analysis, technical aspects, and marketing areas. This paper focuses on maximizing the benefits for development in a sustainable manner toward the producers of traditional processed fishery products and critical key information factors that can be used for the next related research.

16.4 Results

Producers of fishery products in the study site have several characteristics, such as households and individual business with low education level, not having a business license; the equipment and production process are simple with family labors; the marketing area is local, not segmented, upon order demand; the conditions of competition lead to a perfectly competitive market; market expansion opportunities are limited, not having specific marketing strategies; and direct buyers come to the place of production with cash payment system. Thus, the results are in line with some literatures; this type of processing was performed by both men and women in households. The sorting of the fish by species on the basis of freshness and physical damage was done manually. Women played a prominent role in the postharvest activities (Heruwati 2001), and the different fish species were seen for different preparation techniques (Ikeanyionwu 2006). Furthermore, discounted financial indicators are used to provide a notion on costs incurred by different processed fishery products as well as revenues and profits they earn in return, as shown in Table 16.1. By comparing revenues with costs, the analysis reveals how many different actors earn from their businesses.

As shown in Table 16.1, all the processed fishery products are feasible but dried fish has the shortest payback period, the lowest risk compared, and the highest minimum profit limit than the others. Based on this analysis, we can draw some issues related to the empirical study (Kaplinsky and Morris 2001). First, in mapping the distribution of income, we focus on profits; the greater the barriers to entry, the higher the level of profitability. So, profitability is an important window to apprehend the pattern of returns in production networks. The second is the capacity

Table 16.1 NPV, net B/C, IRR, payback period, profitability, and risk analysis

No	Processed fisheries products	NPV (Rp)	Net B/C	IRR (%)	Payback period (year)	Minimum profit limit (Rp)	Coefficient variance
1.	Salted fish	356,306	2.66	24.64	2.01	1,194,846	2.6
2.	Dried fish	547,613	1.78	30.36	0.78	1,360,067	1.8
3.	Fish crackers	644,375	1.29	19.99	1.67	281,548	11.9
4.	Terasi	676,709	0.17	20.00	3.38	684,485	2.5

which individual producers have to upgrade their operations and thus launch themselves onto a path of sustainable income growth. Understanding the determinants of income distribution requires a focus of barriers to entry. Where levels of competition are high, incomes are under threat. The only way in which income growth can be sustained is through an enduring barrier to entry or by the firm, the region, or the country developing the dynamic capability to systematically move of activities in which high barriers to entry prevail. Third, it describes the system of innovation – the network of institutions which supports economic actors. What they do impinges on the competitive performance of firms and groups of firms, and is also subject to the support and regulation provided by governments. Thus, there is a wide consensus that the private sector has a crucial role in achieving pro-poor growth, and the key challenge is to integrate public objectives into private business operations (OECD 2008). The government is urged to provide and to assist through supporting services and implementing a series of measures, with a view to improving the industry's competitiveness and efficiency, and assisting fish farmers to cope with the ever-changing operating environment (The Committee on Sustainable Fishery (CSF) 2010).

Someway, fish processed in various ways will attract people to consume more fish. Even though there are uncertainties about the future of small-scale fishery industries, their output only accounts for a very small percentage of GDP; its sustainable development should not be considered solely from an economic perspective. Generally, by taking into account the local fishery situations as well as various contributions of the small-scale industry to the economy and society, we believe that the government could promote the sustainable manner by the modernization of sustainable modes of operation and fishery resources which continue to be developed.

16.5 Discussion and Conclusions

This study shows that traditional processed fishery products are generally produced by fishermen (households) with low education levels, not having a business license, and the capital itself; equipment and production process are relatively simple, internal labor; local marketing areas, not segmented, by order, its market share is very limited; the conditions of competition lead to a perfectly competitive market; market expansion opportunities are limited, not having specific marketing strategies, and direct buyers come to the place of production by cash payment system. Furthermore, small-scale business of processed products is feasible, but it has moderate to high risks. Parts of the risk are raw materials, production processes, price, and market.

Competing policy interests combined with governance failures, administrative capacity constraints, and changing global fish production and consumption patterns have led to mismanagement, degradation, and overexploitation of fisheries in many cases. In order to reverse these trends, short-term questions of employment and profits of fishers must be carefully balanced against longer-term sustainability. Involving a wide range of stakeholders at global, regional, national, and local levels – such as governments, multilateral institutions, the private sector, regional fisheries organizations, and regional banks – has to be seen as a key factor in ensuring that policies are effective, coherent, and sustainable (Martini 2013).

Thus, the government must provide some form of financial support to small-scale businesses of traditional processed fishery products. The type and level of support vary significantly between regions/districts and take the form of management, research, and enforcement services or more targeted direct support such as payments for vessel construction and modernization, income support, and loan guarantees.

Increasing trade opportunities has important impacts on sustainability and nutrition. The governance of local small-scale business of traditionally processed fishery products, however, can render these negative impacts on development. This does not mean that reduced trade is an appropriate response to an underdeveloped fisheries governance system; rather, increased trade opportunities must come in the context of improved capacity to sustainably supply new markets. Trade promotion efforts must be part of an overall plan for sustainable fisheries development.

Furthermore, traditional fisheries products are subject to regulations related to food safety and quality, food hygiene, packaging, traceability, labeling requirements, and intellectual property protection. While these regulations serve an important purpose, they can also become a barrier to trade especially for developing countries that lack the resources to conform to them. Private certification and labeling initiatives can put additional requirements on exporters from developing countries on top of public regulations. Eco-labeling and other types of sustainability, food quality, or legality certification are increasingly required by major buyers especially in international markets. Due to high costs or lack of data availability, compliance with these schemes may be prohibitive for producers in developing countries.

Some studies also confirm the results of prior studies (Caswell and Padberg 1992; Verbeke and Ward 2006) that labels in general are good and potentially market effective sources of information. Consumers displayed the highest use of expiry date, price, fish species, and weight from fish labels, in a similar vein as findings (Pieniak et al. 2007a) in the case of meat. Moreover, they show the strongest interest in a safety guarantee, quality mark, recipes, and health benefits as potential information cues. Food labels are an important source of information for consumers (Pieniak et al. 2007a, b; Wandel 1997). This is an important result and has practical implications for food marketers and policy makers, which entails opportunities to increase fish consumption by providing/promoting that specific kind of information.

The WTO Agreement on Technical Barriers to Trade helps to ensure that regulatory measures, including regulations, standards, testing and certification procedures, do not create unnecessary obstacles to trade. Policy coherence initiatives must go the extra mile by considering and mitigating the impact on developing countries of nontariff measures in OECD countries (OECD 2012).

Development cooperation can help developing countries build capacity for policy development and implementation in a way that will help them formulate and achieve their objectives. Regulatory frameworks are struggling to keep up with growth, leading to delayed growth, pollution, or other unanticipated outcomes. As some countries solve their management problems, they can share this learning with other countries. In this fashion, it may be possible to avoid repeating mistakes.

A prerequisite for a sustainable small-scale business of traditional fisheries products in developing countries is creating favorable conditions to improve the legal, regulatory, and administrative environments. Reducing corruption, developing institutional capacity, and creating an environment supportive of responsible investment are all means by which countries can lay the foundations for the development of their small-scale business sector.

One way to improve institutional capacity is by participating in regional management agreements. Regional cooperation and sharing of information can offer practical benefits. Regional approaches can also help to drive improvements over time by focusing attention on broader management issues rather than short term national interests. Regional fisheries organizations can coordinate and monitor coherence, for example, through negotiations on agreements. They can ensure the implementation of good management principles by defining minimum conditions for access to fishery resources (OECD 2008).

So, it is urged to take advantage of the opportunities provided by the balance between promoting growth and preserving the long-term capacity of the resource or to make accomplishment through measures including separating the small-scale business area related toward traditional fisheries products into zones and subzones and the formation of the producers for each subzone for management practices identified and introduced with the active involvement for decision-making and for the implementation of management measures. So, the lesson for developing countries from these experiences is that prior planning for fisheries growth is needed to avoid unsustainable or undesirable results.

While it is not possible to anticipate every consequence, taking a proactive approach to regulation and management can save costs. Drawing on other national experiences and drawing up development plans can help to prevent "growing now, cleaning up later." These improvements to regulation and zoning resolved most problems, but the cost of the transition resulted in many firms without sufficient capital being taken over by larger companies (OECD 2013).

In the past, many efforts to speed development ignored the inherent constraints of a sector depending on a limited renewable resource. Surges in capacity led to depleted stocks and damaged local environments. Infrastructure improvements and market liberalization removed barriers to growth that helped to conserve fish stocks by limiting sales. While poor infrastructure is no substitute for good management, development efforts must do a better job of reflecting the unique features of fisheries that do not fit into a standard development model.

In all countries, irrespective of income level, an inclusive approach to sustainable management of natural fisheries resources will lead to broader and more sustainable benefits than one focusing on the short-term interests of fishers. Policy coherence for development cuts across several policy domains and requires effective communication and coordination – not only among government bodies within a country but also among countries. It involves both ensuring that objectives do not conflict and that policies in different domains are mutually supportive to create win-win situations.

Today, however, in an increasingly interconnected world, it is clear that policy coherence also requires paying due attention to other issues as well, notably good governance and the environment. Internationally, regional coalition building can help to establish institutions that deliver improved governance. The government must have a clear understanding of their own policies in terms of priorities, strategies, objectives, and planning.

That individual countries act to put in place sound management policies is a necessary first step, but it is not sufficient to secure the future. Policy coherence for development requires international action and a commitment to multilateralism to create conditions for long-term sustainable management of small-scale business of traditional fisheries products that cross and span global borders. Acting on this recognition of shared interests and responsibilities requires concrete mechanisms such as international frameworks, treaties, and agreements. Much progress has already been made to establish rules and best practices for fishing that apply to everyone, but more needs to be done to put the capacity in place to enforce those rules, to adopt those practices, and to invest in understanding and measuring our shared global business resources.

Strong growth in fishery sectors has already contributed to increased prosperity in developing countries, but its impact on local environments and communities continues to be a concern. Consumers in all countries can help by insisting on products that are responsibly produced and sustainable. Governments can help by aiding producers to meet the highest standards of quality and responsibility for their products. Governments and private industry could find the best technologies, practices, regulation, and management to ensure that growth continues and is sustainable for the long term. Sharing experience in this regard can help to speed the development of this still young sector.

Of particular importance, it is the role of information sharing and knowledge transfer. The availability of correct and consistent information is a key element to define the policies of development needed to ensure economic viability and sustainability of natural resources. An effective system of information and acquisition of data and activities is necessary for the development of rational and sustainable small-scale fisheries in an integrated approach around the Mediterranean and Black Sea. A lot of information on various aspects of small-scale fisheries often escapes or is difficult to obtain or is dispersed among different institutions and organizations at the regional, provincial, or local level. Actually, if structural and technical attributes of the small-scale fisheries are difficult to census due to the high variability linked to multi-specificity and multi-gear characteristics, elements associated to distribution channels, marketing tools, value chain, cooperation, and quality of life issues are even harder to survey. A better understanding of what is currently happening in the basin is fundamental for an efficient and coherent development policy in a multilevel governance and interregional coordination.

We believe that increased responsibility of fishermen organizations in comanagement objectives (also through new forms of territorial rights) is an important step for a sustainable development of small-scale fisheries. However, this technical-productive approach will be probably insufficient to guarantee economic competitiveness of small enterprises without common marketing strategies, and forms of vertical and horizontal cooperation.

In order to coordinate institutional development strategies and local initiatives, public interventions should provide adequate extensions and advisory services for supporting the governance and development of small-scale fisheries processed products. Therefore, the development and support of small-scale business, even if following different legislative paths and conceptual frameworks, should follow a common operational approach to converge toward goals and shared solutions to guarantee successful local strategies in the coastal areas. A common development policy for processed products of small-scale enterprises that sees the participation of all areas would surely and concretely facilitate overcoming difficulties and strengthen the support tools within the different coastal areas.

References

Akabueze B (2002) Financing small and medium enterprises (SMES): the small and medium industries equity investment scheme (SMIEIS) option. N.C Press, Lagos

Al-Jufaili MS, Opara LU (2006) Status of fishery post-harvest industry in the Sultanate of Oman: part 1. Handling and marketing system of fresh fish. J Fish Int 1(2–4):44–49b

Béné C (2006) Small-scale fisheries: assessing their contribution to rural livelihoods in developing countries, FAO fisheries circular, vol 1008. Food and Agriculture Organization of the United Nations, Rome

Béné C, Hersoug B, Allison EH (2010) Not by rent alone: analyzing the pro-poor functions of small-scale fisheries in developing countries. Dev Policy Rev 28(3):325–358

Caswell JA, Padberg DI (1992) Toward a more comprehensive theory of food labels. Am J Agric Econ 74(2):460–468

CFS (2014) Sustainable fishery and aquaculture for food security and nutrition. A report by the High Level Panel of Experts on Food Security and Nutrition. CFS, Rome

Dylick T, Hockerts K (2002) Beyond the business case for corporate sustainability. Bus Strateg Environ 11(2):130–141

FAO (1999) Guidelines for the routine collection of capture fishery data. Prepared at the FAO/DANIDA Expert Consultation. Bangkok, Thailand, 18-30 May 1998. FAO Fisheries Technical Paper. No. 382. Rome, FAO. 1999. 113p. http://www.fao.org/3/a-x2465e.pdf

FAO (2013) Voluntary guidelines for securing sustainable small-scale fisheries in the context of food security and poverty eradication, Rome. 20–24 May 2013

Golub S, Varma A (2014) Fishing exports and economic development of least developed countries: Bangladesh, Cambodia, Comoros, Sierra Leone and Uganda. Swarthmore College. http://www.swarthmore.edu/sites/default/files/assets/documents/user_profiles/sgolub1/UNCTAD.fisheries.final.pdf

Heruwati ES (2001) Traditionally fish processing. Products. Processing Research Centre and Socio-Economic Marine and Fishery. In: Directorate General of Capture Fishery. http://www.pustaka-deptan.go.id

IFREMER coord. (2007) Small-scale coastal fisheries in Europe. Final report, Project No FISH/2005/10. http://ec.europa.eu/fisheries/documentation/studies/small_scale_coastal_fisheries_en.pdf

Ighoroje EJ, Oboro OG (2011) Financing small scale business enterprises in Nigeria: a review of the problems and the way forward. Int J Econ Dev Res Invest 2(3):134–137

Ikeanyionwu C (2006) Entrepreneurship skills for effective management of small scale business in Onitsha, M.sc thesis. Department of Vocational and Adult Education NnamdiAzikiwe University, Nigeria

Kadiri IB (2012) Small and medium scale enterprises and employment generation in Nigeria: the role of finance. Kuwait Chap Arab J Bus Manag Rev 1(9):79–93

Kaplinsky R, Morris M (2001) A handbook for value chain research prepared for the IDRC. http://asiandrivers.open.ac.uk/documents/Value_chain_Handbook_RKMM_Nov_2001.pdf

Köse S (2010) Evaluation of seafood safety health hazards for traditional fish products: preventive measures and monitoring issues. Turk J Fish Aquat Sci 10:139–1605

Macfadyen G, Salz P, Cappell R (2011) Characteristics of small-scale coastal fisheries in Europe. Study IP/B/PECH/IC/2010-158. http://www.europarl.europa.eu/committees/en/studies down load.html?languageDocument=EN&file=49091

Malorgio G, Mulazzani L, Bari C (2014) Enhancing small-scale fisheries value chains in the Mediterranean and Black Sea. General Fisheries Commissions for the Mediterranean. http://www.ssfsymposium.org/Documents/FullVersion/BPIV_Part2.pdf

Martini, R (2013) Fishing for Tomorrow: Managing fisheries for sustainable development. Coherence for Development. OECD. Issue 2. December 2013

OECD (2008) Fishing for coherence in West Africa. OECD, Paris

OECD (2012) The OECD handbook for fisheries managers: principles and practice of policy design. TAD/FI(2012)7/FINAL. Trade and Agriculture Directorate. OECD, Paris

OECD (2013) Green growth and aquaculture. OECD, Paris

Onakoya ABO, Fasanya IO, Abdulrahma HD (2012) Small and medium scale enterprises financing and economic growth in Nigeria. Eur J Bus Manag 5(4):130–136

Onukwuli AG, Akam UG, Onwuka EM (2014) Challenges of small scale industries in sustainable development in Nigeria. J Bus Manag 16(5. Ver. I):19–25. www.iosrjournals.org

Pieniak Z, Verbeke W, Vermeir I, Brunsø K, Olsen SO (2007a) Consumer interest in fish information and labeling: exploratory insights. J Int Food Agribus Market 19(2/3):117–141

Pieniak Z, Verbeke W, Scholderer J, Brunsø K, Olsen SO (2007b) European consumers' use of and trust in information sources about fish. Food Qual Prefer 18:1050–1063

Sacchi J (2011) Analysis of economic activities in the Mediterranean: fishery and aquaculture sectors. Plan Bleu. http://planbleu.org/sites/default/files/publications/analyse_activites_econ_pecheen.pdf

Tawari CC, Abowei JFN (2011) Traditional fish handling and preservation in Nigeria. Asian J Agric Sci 3(6):427–436

The Center for Data Statistic and Information (2014) Marine and fishery in figures 2014. Ministry of Marine and Fishery, Jakarta

The Committee on Sustainable Fishery (CSF) (2010) Report of the Committee on Sustainable Fishery. https://www.afcd.gov.hk/english/fishery/fish_cap/fish_cap_con/files/common/CSF_WP_10_01_2013Eng.pdf

Verbeke W, Ward R (2006) Consumer interest in information cues denoting quality, traceability and origin: an application of ordered profit models to beef labels. Food Qual Prefer 17 (6):453–467

Wals A (2009) Learning for a sustainable world review of contests and structures for education for sustainable development. UNESCO, Paris

Wandel M (1997) Food labeling from a consumer perspective. Br Food J 99(6):212–220

World Bank (2006) Nigeria, poverty in the midst of plenty: the challenge of growth with inclusion. World Bank, Washington, DC

WSSD (2002) The concept of sustainable development. www.ilo.org/public/english/llra/document/congress/Regional/lagos2011/5th session/session 5a/smellscaleent.pdt

Chapter 17
Obstacles to Achieving Food Security: The Failure of Rice Sector Policy and Its Impact on Peasant Deprivation in the Era of Political Reform in Indonesia

Sunardi Purwaatmoko

Abstract The failure of rice sector policy in the era of democratic political reform in Indonesia is related to the interests of political and business powers backed by International Finance Institutions and developed Western countries. The actors attempted to withhold the liberalization policy because doing so would benefit them. The limited government intervention in the paddy and rice market through the decline of the role and function of the National Food Logistic Agency (BULOG) as Public Service Obligation (PSO) prevented the price stabilization of paddy and rice. Rice traders and distributors were able to control the domestic price market of paddy and rice by creating a high price disparity between the market prices of paddy and rice. The storage of paddy and rice purchased by private rice traders and distributors from peasant producers led to high disparity between domestic and global rice prices. The higher price of domestic rice accompanied with the low-level import tariff made it easier for rice importers to import and access the market. Consequently, the failure of de-liberalization policy resulted in disadvantages for poor people and peasant producers.

Abbreviations

AGRA	Aliansi Gerakan Reforma Agraria (Alliance for Agrarian Momement Reform)
API	Aliansi Petani Indonesia (Indonesian Peasant Alliance)
BAPPENAS	Badan Perencanaan Pembangunan Nasional (National Development Planning Agency)

S. Purwaatmoko (✉)
Department of International Relations, The Faculty of Social and Political Sciences, The University of Jember, Jalan Kalimantan No. 37, Jember 68121, East Java, Indonesia
e-mail: sunardiatmk@yahoo.com

B. McLellan (ed.), *Sustainable Future for Human Security*,
https://doi.org/10.1007/978-981-10-5430-3_17

BPS	Badan Pusat Statistik (Centre for Statistics Agency)
BULOG	Badan Urusan Logistik (National Food Logistic Agency)
DPR	Dewan Perwakilan Rakyat (Parliament)
DTI	Dewan Tani Indonesia (Indonesian Peasant Council)
EU	Uni Eropa (European Union)
GERINDRA	Partai Gerakan Rakyat Indonesia Raya (Greater Indonesian People Movement Party)
GKG	Gabah Kering Giling (dried paddy on miller)
GKP	Gabah Kering Panen (dried paddy on farm)
GOLKAR	Golongan Karya (Functional Group Party)
HANURA	Hati Nurani Rakyat Party (People Approval Party)
HBE	Harga Beras Eceran (retail price of rice)
HDG	Harga Dasar Gabah (floor price of paddy)
HDGB	Harga Dasar Gabah dan Beras (floor price of paddy and rice)
HDPP	Harga Dasar Pembelian Pemerintah (government procurement of floor price)
HGTP	Harga Gabah Tingkat Petani (price of paddy at the level of peasant)
HPP	Harga Pembelian Pemerintah (government procurement price)
IFIs	Lembaga-Lembaga Finansial Internasional (International Finance Institutions)
IGJ	Institute for Global Justice
IMF	International Monetary Fund (Dana Moneter Internasional)
INPRES	Instruksi Presiden (Presidential Decree)
KKP	Kredit Ketahanan Pangan (Food Resilience Credit)
KPA	Konsorsium Pembaruan Agraria (Consortium for Agrarian Reform)
KUD	Koperasi Unit Desa (Village Unit Cooperative)
LKBI	Likuiditas Kredit Bank Indonesia (Bank Indonesia Liquidity Credit)
MH	Musim Hujan (wet season)
MK1	Musim Kering 1 (first dry season)
MK2	Musim Kering 2 (second dry season)
NGOs	Nongovernmental Organizations (Organisasi Non-Pemerintah)
OP	Operasi Pasar (market operation)
PAN	Partai Amanat Nasional (National Speech Party)
PBB	Partai Bulan Bintang (Moon and Star Party)
PDIP	Partai Demokrasi Indonesia (Perjuangan Indonesian Democratic Party of Struggle)
PIBC	Cipinang Rice Market Centre (Pasar Induk Beras Cipinang)
PKB	Partai Kebangkitan Bangsa (National Awakening Party)
PKS	Partai Keadilan Sejahtera (Justice and Welfare Party)
PPSE	Pusat Penelitian Sosial Ekonomi (Center for Social and Economic Research)
PSO	Public Service Obligation (Tanggung Jawab Pelayanan Publik)

RASKIN	Beras Miskin (rice for poor people)
REPELITA	Rencana Pembangunan Lima Tahun (five-year development plan)
RMU	Rice Milling Units (Unit Penggilingan Padi)
Rp.	Indonesian Rupiah
RT	Rumah Tangga (household)
SBY	Susilo Bambang Yudhoyono
SPI	Serikat Petani Indonesia (Indonesian Peasant Union)
WAMTI	Wahana Masyarakat Tani dan Nelayan Indonesia (Indonesian Peasant and Fisher Organisation)

17.1 Introduction

In the era of political democratization beginning from Abdurrahman Wahid (2000), Megawati (2001) and Yudhoyono (2004), strong government intervention in the domestic markets of agricultural products was considered necessary since the liberalization policy did not guarantee income security for the small-scale farmers and poor people. The justification provided for the government intervention was based on the principle that an institutional structure would be able to align the agricultural sector with the government-desired targets based on strategic and social weights. As advanced democratic countries, the United States, Japan and European Union have fixed a high import tariff and price stabilization policy on agricultural products. For example, the European Union (EU)[1] specified a rice import tariff of 211 Euro per metric ton, about 2000 Rp (Indonesian Rupiahs) per kilograms under the assumption that the value of the Euro was 9500 Indonesian Rupiahs, while Japan and the United States fixed import tariffs higher than the European Union. The tariff on imports in Japan was 402 yen per kilogram or 30.150 Rupiahs, with the value of the Yen as 75 Indonesian Rupiahs. The tariff on imports in the United States was 2.1 US dollar (18.900 Indonesian Rupiahs) per kilogram, with the value of the US dollar as 9000 Indonesian Rupiahs. In contrast to these states, Indonesia only imposed a tariff on rice imports of 430 Indonesian Rupiahs per kilogram.[2] During the period when Indonesia was governed by Suharto, although under an authoritarian political regime, the stabilization efforts of the National Food Logistic Agency (BULOG) paid very high dividends in fostering faster economic growth during REPELITA I and II.[3] The production of the agricultural sector has always faced short-term fluctuations because it is affected by natural factors prone to changes such as land, seasons, weather conditions and, at times, plant diseases. The crops of agricultural products are uncertain, and external shocks affecting the balance of the free market frequently influence the price level of crops. The supply of agricultural products was not responsive to the changes of market conditions in

[1]All expansions are available at the above (after the abstract).

[2]*Kompas*, 27/2/2003.

[3]Timmer, Peter, (2004), Food security in Indonesia: current challenges and the long-run outlook. *Centre for Global Development*, p 14

the short term.[4] Hence, theoretically, agricultural sector should not be subjected to the free market mechanism in Indonesia because if it were applied in the country, it would have a negative effect on the small-scale peasant producers and the low-income consumers.

To achieve price stability for paddy and rice, the government must fix the floor price of agricultural products. The aim of this price stabilization policy was to protect both the peasant producers and consumers. The proposed solution for price stabilization was buying agricultural products at the minimum guaranteed price and selling it to consumers when the market price level was above the ceiling price. The government built a national food institution for buying and selling agricultural products from peasant producers to maintain price stability. The purpose of the price stabilization policy was to ensure that private business actors such as big Rice Milling Units (RMU) and distributors would not create an oligopoly in the rice market. The supporting instruments were tariffs and import restrictions. In advanced industrial countries with high-income per capita where the income of peasant families is much lower than that of the income of people working in the industrial sector, the government was normally pursuing higher protection policies for agricultural products. The level of protection depended on the level of economic advancement of certain states. In the advanced industrial countries, the higher the income of the people working in the industrial sector, the higher was the protection offered to the peasant producers. These countries used protection measures to alleviate the income gap between peasant producers and the people working in the industrial sector. Hence, based on the findings of liberal theorists, the agricultural sector should not be subjected to the free market mechanism.

Let us now compare the situation of Indonesia in the era of political democratization with that of advanced countries. The policy of liberalization in the rice sector adopted in the era of political democratization had negative effect on the small-scale peasants living in the villages and poor people. Unlike the situation of the highly industrialized countries such as the United States, Japan, South Korea, the European Union and Suharto regime before economic crisis, under the power of IMF, World Bank and the supporting actors, the democratic reformed government of Indonesia in 2000 continued to undertake policies towards rice sector liberalization.

Beginning in 1997, the Suharto government suffered a severe economic crisis which forced the government to call for IMF and the Western States to give economic support to restore the economy. For this reason, in the post-economic crisis, the rice policy in Indonesia was substantially reformed because IMF and the Western States wanted the government of Indonesia to accept Structural Adjustment Packages (SAP), liberalization, as a prerequisite for the economic aid. The implications of the SAP agreement between IMF and the government of Indonesia in 1998, began to change substantially the rice sector policy, liberalize the domestic market of rice and fertilizer and rice import and reform BULOG.[5]

---

[4]Ibid, p.168.

[5]Setiawan, Boni, 2006, Ekonomi Pasar yang Neo-liberalistik Versus Ekonomi Berkeadilan Sosial, *Forum Komunikasi Partai Politik dan Politisi Untuk Reformasi di DPR RI.*

Following the adoption of democratic political order in 1999, the reformist political parties led by Abdurrahman Wahid and then Megawati Sukarnoputri attempted to address the limitations of the liberalization policy. However, following Sukarnoputri's political defeat in the public election in 2004, SBY, the winner of the election, attempted to retain the rice policy liberalization. As a result, the interests of small-scale peasants and poor people became marginalized.

In the era of political democratization, political parties and interest groups flourished.[6] Twenty-four political parties joined in the election in 1999.[7] On the level of civil society, peasant organizations and NGOs emerged, SPI, API, Petani Mandiri (Independent Peasant), KPA, AGRA, DTI, WAMTI, KRKP and IGJ. They instigated public protests against domestic rice market liberalization and rice imports through public demonstrations and criticism in the press. Although the Megawati government initially attempted to prevent the spread of rice sector liberalization by banning the rice imports and subsidizing fertilizers for farmers, the SBY government was reluctant to reform the liberal policy because the government was closely linked to the political and business actors dominating the rice sector policymaking process.[8] The question was why the democratic system adopted was not able to provide reasonable income for the small-scale peasants and poor people in the villages. This research aimed to identify the political behaviour of political and business actors and peasant organizations and NGOs involved in the policymaking process in the rice sector in the new democratic government in Indonesia.

17.2 Argument

Antony Down argued that the behaviour of constituents can be directed by legislating to the *median voter*.[9] However, public policy does not always represent the interests of the *median voter*. This argument only materialized in democratic society where behaviour of constituents was restricted by a two-party system. In the multiparty system, the constituents tended to move beyond the *median voter*. Consequently, both public and private service would be available to interest groups not included within the *median voter*.[10] To get constituent votes, politicians needed

[6]*Kompas*, 22/05/2004.

[7]*Kompas*, ibid., *2004*.

[8]Nurruddin, 2009: *Interview*.

[9]Down, Antony, quoted from *Theories of Political Economy*, James A Caporaso and David P. Levine, 1992. P.139–141.

[10]Geddes, Barbara, (1994), *Politician Dillema: Building State Capacity in Latin America*, University of California press, Berkeley/Los Angeles, p 39

a big political machine consisting of political party activists and party workers willing to work for them in return for their financial benefits. They were recruited for delivering gifts to the constituents.[11] For that reason, the political parties were not independent in formulating policy preferences because the only aim they pursue was re-election in the political competition. They had rational reasons to deliver resources under their authority distributed to individuals as gifts to sustain their political carreer instead of giving public service.[12]

To support the argument, Dahl and other pluralists suggest that in America, as a democratic country, power and authority had been distributed among officials of civil servants, private individuals and interest groups. The structure of power was segmented and unorganized without a clear pattern of hierarchy.[13] In contrast to this, the democratic government system adopted by Indonesia showed different case. In Indonesia, power and authority were not yet distributed among officials of civil servant, private individuals and interest groups. Peasant organizations and NGOs, as elements of civil society in Indonesia, were unable to influence policymaking and implementation. Consequently, the policymaking process and implementation were dominated by political oligarchies.

The process of democratization in Indonesia progressed without abandoning the elements of the old political oligarchy. The adapted multiparty system gave rise to fierce political competition among political parties to achieve political success at the cost of their public service responsibilities, because they were trapped in rent-seeking practices involving the misuse of government resources. The reformist political parties were split into independent and pragmatic parties (PAN, PKB, PBB and PKS) attempting to build political coalitions with the old political oligarchy. Although the old oligarchy had also become divided into new circles of political powers (GOLKAR, DEMOKRAT, HANURA and later GERINDRA), their political powers were nevertheless dominant in the policymaking process of the rice sector. PDIP as a reformist political party was forced to stay alone as an opposition party.

The political coalitions between the elements of power of the old New Order regime oligarchy and the pragmatist political parties supported by international finance institutions and debtor countries had the political capacity to prevent the political participation of the peasant organizations and NGOs. The political coalition was therefore an obstacle to achieving their desired food security policy.

[11]Ibid, p:40.

[12]Ibid, p:41.

[13]Chilcote, Ronald H., (1981), *Theories of comparative politics: the search for paradigm.* Westview Pres Inc., p 353

17.3 The Failure of Government Policy and Its Impact on Farmer Income and Rice Consumers

The adoption of liberalization policy in Indonesia began in 1998 when the IMF urged the government to reform the rice policy by reducing the function and role of BULOG. Based on the Frank Ellis survey by the National Development Planning Agency (BAPPENAS), the government decided to reform BULOG. Ellis, an academic expert of University of East Anglia, was recruited by the government as an economic adviser. He argued that the Indonesian peasant was the net consumer of rice. For that reason, the scale of rice procurement operations could be reduced. The price stabilization of rice at the level of peasant producers should be concentrated in specific regions and seasons. The Ellis argument was further strengthened by the BAPPENAS analysis which estimated the BULOG operation cost by conducting simulations of rice storage and price stabilization functions. BAPPENAS found that the government's financial burden could be reduced by reducing the volume of BULOG's rice procurement and lowering the rice stock level.[14]

Based on the Ellis survey and BAPPENAS simulation, it was decided that BULOG as STE needed to be reformed. The level of rice stock needed to be reduced to alleviate government state-owned trading enterprise spending, and BULOG was required to focus on the supply of rice for poor families and other budgeted groups. The National Rice Stock was austerely set at about 1 million to 1.25 million ton or slightly more under the management of BULOG. The stock was considered adequate when the rice stored in BULOG reached 2 million ton. The paddy and rice floor prices were not decided by BULOG but by the government. The quantity of government procurement of rice was curtailed at approximately 2 million ton. The domestic rice procurement after the BULOG reform was thus reduced by 50%. In the New Order regime, the government procurement of paddy and rice reached up to 4 million ton. Because of this reason, the Suharto regime did not encounter any difficulties in formulating a plan for the price stabilization of paddy and rice. Paasch et al. explained the difference between HDGP and HDPP policy options.[15] The data in Table 17.1 shows that the first

[14]Yonekura, Hitoshi, (2005) Institutional Reform in Indonesia's Food Security Sector: The Transformation of BULOG Into Public Corporation. *Developing Economies XLIII-1 (2005)*:313.

[15]In the era of Suharto, BULOG was entrusted to stabilize the domestic rice price through LKBI soft loan credit mechanism and had been successful in implementing the stabilization of rice price, relative to the democratic government in the era of political reform because in this era LKBI was abolished and changed into government budget used for stabilization policy of rice price allocated to BULOG. In the era of Suharto, a nondepartmental institution, BULOG, had full authority to carry out HDGB policy, while in the era of political reform, the policy of HDPP was determined by the government.

Table 17.1 Floor price policy (HDG) and government procurement price policy for paddy

Description	HDGB Policy	HDPP Policy
Policy objectives	Maintain price at or above a minimum floor level (HDG) throughout the year	Help to buffer price of paddy, particularly during harvest, to avoid unacceptable low price
Policy instruments	Purchasing of paddy at the guaranteed minimum level until the market price rises above that level (no limitation of purchasing volume)	Purchasing of paddy up to a maximum volume at a guaranteed minimum price (no mandate to keep market price of paddy above the minimum price)
Supporting instruments	Tariff and import limitation	Tariff and import limitation
Policy effectiveness	Effective to keep the marketplace price of paddy above the stipulated level	Effective to stabilize prices outside the high season; not as effective during harvest because of the too limited volume that comes under this measure
Costs of policy	Approximately double the cost of the second option because it was usually necessary to buy about 4 million metric tons (approximately Rp 16 trillion) of rice during the peak harvesting season to keep the price up	Currently, the purchased volume of paddy is 2 million metric tons (around Rp 8 trillion)
Political accountability	Government is obliged	Government is not obliged

Source: Paasch et al. (2007)

policy option was better than the second but caused the government a greater financial burden.[16]

The following section discusses how the term 'floor price of paddy and rice (HDGB)' transformed into 'the Government Procurement of Floor Price (HDPP)'. After the reform, the term 'floor price of paddy and rice (HDGB)' was transformed into 'the government procurement of floor price (HDPP)'. Based on the scheme, BULOG was responsible for buffering the price of paddy to avoid the downturn of paddy price at the level of small-scale peasant producers but was not responsible to maintain the market price of paddy above the minimum guaranteed price because BULOG was only in charge of purchasing up to only a certain amount at the guaranteed price specified by the government. In exchange, the government was responsible for providing financial support for the delivery of rice to poor people, for market operation when the domestic rice price escalated and affected the consumers and for people harmed by natural disasters. The aims of the government policy were to reduce the negative effects, for both the peasant producers and the

[16]Paasch, Armin et al. (2007), *Kebijakan Perdagangan dan Kelaparan: Dampak Liberalisasi perdagangan Terhadap Hak atas Pangan Komunitas Petani Padi di Ghana, Honduras, dan Indonesia*, Ecumenical Advocacy Alliance, Geneva.

consumers, of the sudden drops of paddy and rice price at the level of peasant producers in the growing season and the escalation of consumer rice price during dry seasons. However, the effectiveness of the government policy depended on the accessibility of government spending on rice for the poor people, market operation (OP) and occurrence of natural disasters. The government failure to supply and allocate these three government-spending components destabilized the balance of power of commercial transactions between the peasant producers and the rice consumers, causing a negative impact on the stability of paddy and rice prices.

The government procurement price of paddy and rice (HDPP) policy was effectively implemented outside the high seasons and was not as effective during the harvest season because of the limited volume of crops that was being produced. Consistent with the academic foundation constructed by Ellis and BAPPENAS, in 2002 up to SBY government in 2010, the volume of purchasing paddy and rice was restricted to 2 million ton as a result of the austerity program launched by the government who worked with the IMF during the economic crisis.

The transformation in terminology from HDG into HDPG indicated that the government no longer had formal responsibility to guarantee the floor price at the minimum guaranteed price at the level of the peasant producer. BULOG did not have the responsibility to defend the floor price of paddy (HDG) but only bought paddy to suit its own needs. The price of paddy dropping instantly in the market under the price level of HDPG was beyond the responsibility of BULOG.[17] The HDPP policy was conceptually different from the HDPG policy. In the HDPG policy, BULOG had been completely in charge of stabilizing both producer and consumer price through public purchasing of paddy and rice and market operation (OP). BULOG was allowed to buy as much paddy and rice from the peasants as they needed through the soft credit mechanism of Bank Indonesia Liquidity Credit (LKBI) until the market price rose above the guaranteed minimum level. Hence, the scale of procurement operation of paddy and rice was approximately double the cost of the HDPP policy option because it was usually necessary to buy approximately four million metric tons of rice during the peak harvesting season to keep the price up.

The BULOG reform indicated that the government intervention in the rice market had significantly reduced, and most of the domestic experts questioned whether the decision was appropriate. Mostly, the domestic agronomists doubted the truth of the Ellis argument that the small peasants in Indonesia were net consumers. A survey by Sumaryanto et al. (2002) indicated that the Indonesian peasant was actually a net producer. The peasant producers cultivated rice during the wet season (MH), first dry season (MK1), and second dry season (MK2). Each farmer family cultivated 0.35, 0.33, and 0.23 hectare of rice field on average. Information about the amount of land cultivated was available in Indonesia, particularly for paddy cultivation. The per capita consumption of rice was

[17]Simatupang, Pantjar, *Suara Pembaharuan*, 28/1/2003.

107 kilograms/cap/year. The amount of consumption of rice did not include indirect rice consumption such as eating at restaurants, feasts, and workplace.

Field research conducted by Sumaryanto in 1999/2000 indicated that the total rice production was approximately 462 kg/capita/year on average. When the production of rice was reduced owing to peasant family consumption, a marketable surplus of approximately 354 Kg/cap/year continued to remain. The marketable surplus in the wet season was higher than the first dry season (MK1), but negative in the second dry season (MK 2). Many peasants did not plant paddy because of the shortage of irrigation water in the dry season. Accordingly, the total size of cultivated land reduced. Thus, it was not accurate to assume that the small-scale peasant producers in Indonesia were net consumers throughout the year. They were net producers in the first two seasons (MH and MK1) and net consumers in the second dry season (MK2). The assumption that the small-scale peasant was a net consumer was not supported by empirical data. Generally, the peasants sold their surplus production in the wet season and first dry season (MK1).[18] Hence, based on Sumaryanto's findings, the massive drop in the government procurement quantity of paddy and rice by BULOG had given the rice traders and distributors an opportunity to store the production surplus of the paddy and rice produced by the small-scale peasant producers because the peasants had high expectation about selling their paddy and rice production. Consequently, the drastic decline of the government procurement of paddy and rice gave rise to market anomalies by handing over the market power to monopolists/oligopolists. The rice market analysis explained below proves that the Ellis survey and BAPPENAS simulation were not empirically supported. The decreased quantity of government procurement of paddy and rice was unable to sustain the price stability and had a negative impact on both the peasant producers and poor consumers.

Table 17.2 differentiates the quantities of the public procurement of rice, rice for the poor and rice import from the year 2000 to 2007. Referring to the data, the annual public procurement of rice, rice delivered to the poor (RASKIN) and rice import were on average 1,884,525 tons, 1,814,486 tons and 1,414,202 tons, respectively. With the average quantity of 1,884,525 ton of annual public procurement and the average annual delivery of 1,814,486 tons of rice for the poor people, the domestic rice market was inelastic although the annual import of rice was 1,414,202 tons on average, and market operation (OP) had been launched by the government when the domestic rice production increased. It proved that the annual total of public procurement of domestic rice and rice import from overseas countries had no capacity to combat the dominant power of the domestic rice traders and distributors. The Ellis Survey and BAPPENAS simulation that had been used as an academic foundation for the BULOG reform were not adequately connected to or supported by the actual empirical data.

[18]Sawit, M. Husein, (2007), Usulan Kebijakan Beras Dari Bank Dunia: Resep Yang Keliru, *Pusat Analisis Sosial Ekonomi dan Kebijakan Pertanian*, Jl. A. Yani No. 70 Bogor 16161 (article), Sawit, pp 195–196.

Table 17.2 The volume of rice procurement, rice for poor and rice import

Year	Rice procurement (million ton)	Rice for poor people (million ton)	Rice import (million ton)
2000	2,174,807	1,350,000	1,512,000
2001	2,018,338	1,500,000	1,404,051
2002	2,131,608	2,235,000	3,707,037
2003	2,008,954	2,023,664	2,775,328
2004	2,096,610	2,060,198	632,000
2005	1,529,718	1,991,131	304,000
2006	1,350,181	1,624,089	840,000
2007	1,765,987	1,731,805	1,500,000
Average	1,884,525	1,814,486	1,414,202

Source: BPS and TRR (author's analysis)

Referring to the findings of Sumaryanto et al. (2002), the volume of public purchasing of domestic paddy and rice needed to be improved until the market price increased above the guaranteed minimum level. The production surplus of small peasants in Indonesia was stored in BULOG warehouses in the harvest season and was sold to the market in the hard seasons when the consumer rice price escalated. In contrast, following the decrease of the government purchasing of domestic paddy and rice, the market price of paddy and rice in Indonesia came under economic control of rice importers and domestic distributors possessing modern rice milling units, paddy dryers and vast warehouses. The domestic distributors possessing these equipment influenced the market price. The rice importers and distributors had built a strong business relation with the pragmatist political parties and bureaucrats involved in the government administration. Consequently, the bargaining position of the peasant producers was weak because the peasant organizations and NGOs did not have the political capacity to support the peasants. Only a few political parties politically supported them, namely, PDIP and PKS, because most political parties created in the era of democratization were pragmatic and not interested in protecting the farmers. Although PKS had been involved in the government coalitions, the party did not agree with the policy of liberalization.

Since 2005–2007, the domestic rice purchasing has decreased below 2 million metric tons, and the delivery of rice for the poor people was approximately the same as the quantity of government procurement, which had an impact on increasing the consumer rice price but reducing the paddy price. Filling the sparse supply of rice in BULOG warehouses would decrease the domestic consumer rice price if BULOG were to launch the policy for domestic paddy and rice procurement. However, BULOG preferred to import rice from overseas countries rather than purchase domestic paddy and rice because it was more profitable, although both the peasant producer and the consumer were offended. Political protests were launched against the rice import policy by the ideologist political party, PDIP, supported by peasant organizations and NGOs linked to PDIP and the heads of local government in 2005 and 2006. However, the SBY government was reluctant to purchase domestic paddy and rice and preferred to work with BULOG which was supported by high-class rice distributors and rice importers. Finally, the government officially

Table 17.3 The price level of HPP GKG, HGTP, HBE and HBE (PIBC) 1980–2006

Year	Government procurement price of dried paddy on miller (HPP GKG) (RP./kg)	Price of paddy on farm (HGTP) (Rp./kg)	Retail price of rice (HBE) (Rp./kg)	Cipinang Central Market of Rice (PIBC) (Rp./kg)
1980	105	189,32	198,39	186
1981	120	212,16	226,19	205
1982	135	229,61	254,92	226
1983	145	229,61	304,24	275
1984	165	274,69	330,97	286
1985	175	284,81	322,07	254
1986	175	288,59	345,24	333
1987	190	167,27	386,86	388
1988	210	184.73	469,2	387
1989	250	381,62	469,56	404
1990	270	475,48	525,17	430
1991	295	488,68	557.84	452
1992	330	517,47	603,68	696
1993	340	303,7	592,25	722
1994	360	284,05	660,37	900
1995	400	325,83	776,38	1087
1996	450	419,81	880	1185
1997	525	432,75	1064,03	1285
1998	800	498,27	2099,71	2552
1999	1400	933,01	2665,58	2461
2000	1400	1159,43	2424,22	2099
2001	1500	964,72	2537,09	2256
2002	1519	1141,22	2826,06	2678
2003	1700	1255,46	2785,85	2704
2004	1700	1258,32	2850,96	2600
2005	1740	1567,67	3478,87	3046
2006	2280	2094	4378	4077

Source: BPS and Kompas (25/11/ 2006)

began to import rice by breaking the rice import ban stipulated by the Megawati government in 2004. The newly elected SBY government was supported by the old political oligarchy, GOLKAR, and its splinter political party, DEMOKRAT, and was building a political coalition with pragmatic political parties PAN, PKB and PKS. The multiparty system adopted in the era of political reform compelled new political parties to become pragmatic due to high political competition among them. Consequently, under these political conditions, the elements of the old political oligarchy survived.

Data presented in Table 17.3 reveals that since the end of the New Order regime (1998–2006), the price of paddy at the level of the peasant (HGTP) continued to decrease below the level of HPP of dried paddy on miller (GKG). This implied that the level price of HPP was not effective to sustain the price stability of paddy price

Table 17.4 Import of rice from 1998 to 2013

Year	Rice import (BPS)	Rice import (TRR)
1998	2,900,550	6,077,000
1999	4,751,850	4,183,000
2000	1,375,498	1,512,000
2001	644,733	1,404,051
2002	1,805,380	3,707,037
2003	1,428,506	2,775,328
2004	236,867	632,000
2005	189,617	304,000
2006	210,000	840,000
2007	1,500,000	1,500,000
2008	289,274	–
2009	250,276	–
2010	687,583	–
2011	2,744,261	–
2012	1,927,000	–
2013	472, 664	–

Source: BPS and TRR (The Rice Report)

in a manner beneficial to peasant producers and consumers. The increase in price disparity between HGTP and the retail price of rice (HBE) indicated that the volume of the public procurement of paddy and rice on the farm specified by BULOG was unable to sustain the domestic price stability of paddy and rice. The increase in price disparity between rice and paddy indicated that buying and selling transactions of paddy and rice only benefited the rice traders but harmed both the producers and poor consumers. Price drops of paddy on farm (HGTP) occurred, causing the price to remain under the HPP price level.

The data presented in Tables 17.3 and 17.4 reveal an interesting trend. The escalation of rice import volume in 1998–2003 was accompanied by the downturn of domestic rice price in 1998–2000. Although the rice import prohibition specified by the Megawati government in 2004 had resulted in a drop in the import volume in 2004–2006, this policy was unable to significantly increase the price of paddy. Additionally, the consumer price of rice tended to remain above the price of paddy. The trend confirmed that the rice import ban policy stipulated by the Megawati government did not have a positive impact on the economic welfare of the peasant producers and impoverished people. The oligopolistic structure of the domestic market continued to increase the incentives of the rice traders and high-class distributors at the expense of rice consumers. Consequently, not only the peasants and poor consumers but also the middle-class rice consumers were negatively affected because of liberalization policy. The small-scale peasants did not achieve a suitable income from cultivating their land. With income less than US$ 1 per day, it was very difficult for them to bear the expenses of their daily lives and finance their children's education.[19] The number of small-scale peasant families in

[19]Op-cit, 2007, 99–100.

Indonesia amounted to approximately 13.7 million,[20] 75% of whom cultivated 38% of the total land area.[21]

The downturn price of retail price of rice (HBE) was due to the reluctance of BULOG to purchase domestic paddy and rice. The transformation of BULOG as a state-owned trading enterprise (BUMN) justified the limited state intervention in the rice market. In the post-reform period, BULOG was not responsible for purchasing paddy and rice from the peasants although the retail price of paddy (HGTP) had reduced under the level of HPP. BULOG did not have the responsibility to defend the HPP but only bought paddy and rice for its own need. In such a situation, the middlemen, small and medium millers and high-class distributors had an opportunity to reduce the price of paddy and rice produced by the peasants. BULOG had the legitimate power to import rice from overseas countries instead of purchasing domestic paddy and rice because purchasing domestic paddy and rice did not enhance the profit margin. As a result, although the government had stipulated low-priced HPP and the market price of paddy was under the price level of HPP, BULOG preferred to import rice because BULOG as a state-owned trading enterprise was forced to generate profit rather than conduct public service obligation. The current government did not have the legitimacy to compel BULOG to purchase more domestic paddy and rice than BULOG needed to the extent that it was not re-reformed. HPP was not effective in sustaining the price stability of paddy and rice because the market prices were mostly under the price level of HPP. The respondents interviewed by Pratiwi stated that the price level of HPP was unable to support the peasants' interests. Although the formal value of HPP was increasing all the time, its value escalations did not match the production costs, causing the small peasants' income to become even lower.[22]

Since the end of the Suharto regime, the market price disparity between paddy and rice began to increase. Empirical data on the market price disparity show that the value of rice processing (the cost of transforming paddy into rice) and trade was only enjoyed only by rice traders and distributors at the expense of the peasant producers and consumers.[23] The rice market anomaly due to the dominant power of rice traders and distributors not only depressed peasant producers and middle-class consumers in the cities but also the poor people in the villages. Rice import pressures in the inelastic market structure jeopardized the daily lives of the peasant producers and poor people living in the villages. In an inelastic free market, rice traders and distributors had the capacity to augment their profit margin by suspending the increase of peasant producer' prices when the consumers' rice price was escalating. In contrast, rice traders also had the capacity to achieve the same profit margin by hastening the decrease of market price at the level of peasant producers although the consumer price was

[20]*SUSENAS, 2003.*

[21]Departemen Pertanian, *Sensus Pertanian, 2003.*

[22]Pratiwi (2008), *Efektivitas Dan Perumusan Strategi Kebijakan Beras Nasional*, Program Studi Manajemen Agribisnis. Fakultas Pertanian Institut Pertanian, Bogor, p 95.

[23]Arifin, Bustanul, (2007), *Diagnostik Ekonomi Politik Pangan dan Pertanian*, PT Raja Grafindo Persada, Jakarta, p 238.

Table 17.5 Price disparity of domestic and international rice

Year	Domestic rice price (Rp/Kg)	International rice price (Rp/Kg)
1980	198	180
1985	345	221
1990	525	468
1995	776	683
1996	880	778
1997	1063	2181
1998	2099	2117
1999	2655	1576
2000	2215	1251
2001	2450	1762
2002	2842	1572
2003	2759	1542
2004	2795	1946
2005	3332	2116
2006	4378	2424

Source: BPS (2006)

decreasing, causing all the business risks to be diverted to the peasants' burden. Thus, the rice traders had the capacity to isolate peasant producers from the free market because the actual market players were the rice traders and the consumers.[24]

The price escalation of domestic rice and the decrease of paddy price as a result of market price manoeuvring by the domestic rice traders and distributors (Table 17.3) had an impact on speeding the decline of international rice price (Table 17.5), thereby increasing the incentives for domestic and international importers. Consequently, the rice import volume began to increase, and the peasants' interests were in conflict. The data presented in Table 17.5 reveals that beginning from the year 1999 up to 2006, the international rice price was continually decreasing relative to the domestic rice price.

17.4 Pragmatic Political Parties and Business Interests

During the ensuing years after the economic crisis, the rice supply in the domestic market was restricted by rice speculators' networks controlling the lower and middle-class rice distributors. The network had their own warehouses with more storage capacity than those of BULOG. Cartel/oligopolistic practices of rice were integrated with other food products and involved politicians and rice distributors.

[24]Mardiyanto, Sudi; Supriatna, Yana; Agustin, Khoiriyah, Nur, 2005, *Dinamika Pemasaran Gabah dan Beras*, Pusat Analisis Sosial Ekonomi dan Kebijakan Pertanian, Bogor. (article), 2005, p.127.

They were effectively organized and protected by political actors at the government level. The overseas exporters cooperated with public service officers, members of Parliament (DPR), law officials and politicians.[25] The food cartel prices indicated similar trends but were determined by different actors.[26] The rice import contributed to a more complex food security problem. The disparity between international and domestic prices due to domestic oligopolistic practices had increased incentives for rice importers. In addition, the price level of international rice was more competitive relative to the domestic rice because most of the exporting countries protected their domestic farmers.

The business relations between bureaucratic agents, rice traders and distributors were undoubtedly unique.[27] They had built mutually beneficial business relations but predatory ones, in the sense that their business relations negatively affected peasant producers and impoverished people. Since the intervention of the IMF, the soft loan through the Bank Indonesia Liquidity Credit (LKBI) mechanism was removed, causing severe weakening of the BULOG's function in stabilizing domestic rice price. The government did not specify HDG at the level of peasant producers, and KUD was not responsible for buying paddy from the peasants and sold it to BULOG. HPP was determined at the BULOG warehouse level, and BULOG was allowed to work with private rice millers and distributors for the procurement of domestic paddy and rice. As a result, the function of KUD to buy domestic paddy and rice directly from the peasant producers had been abolished. Hence, the peasant producers in the villages lost their protector. The weak village famers did not have other alternative outlets for their paddy and rice productions. Following the elimination of KUD as the sole agent of BULOG in the procurement of paddy and rice, the only market access of rice and paddy for the village famers was the middlemen, small and middle-class traders and high-class distributors. This implies that the government had been allowing the farmers to engage in a lonely fight with them in the free market of rice and paddy without any government protection.

The first group of actors was middlemen, lower-class rice millers, middle-class and high-class rice distributors, importers, exporters and BULOG gaining profits from liberalization of domestic rice market and import policy. The lower-class rice millers were sometimes injured by surges of rice import, but they could cover their loss by mixing the expensive domestic rice with the cheap-imported rice and sold it to the consumers. This political power grouping was strongly supported by vestiges of New Order regime oligarchy which had been separated into poles of political powers after the political reform.[28] However, they were still dominant in the policymaking process. In contrast to this, the second group of actors were peasant producers and poor consumers who had been severely harmed from domestic rice

[25]Daily news, *Rakyat Merdeka*, 18/02/ 2013.

[26]http://www.suarapembaruan.com/home/k...ia-tanah/29802

[27]Sujatmiko, Budiman 2012: *Interview*.

[28]Robinson, Richard and Hadiz, Vedi R, (2004) *Reorganizing power in Indonesia: The Politics of Olygarchy in An Age of Markets*. Rootledge Curzon, London. EC4P4EE.

liberalization policy and surges of rice import. In an oligopolistic market both the peasant producers and the consumers had been debilitated. The poor consumers and the peasant producers were politically supported by reformist political parties, heads of local governments, peasant organizations and NGOs. They also consisted of separated independent political power but temporarily united based on their own political reasons and interests. The second political grouping didn't have political power resources suitable for influencing policymaking process because most of them were the new comer political actor which could be termed as the new political oligarchy.

Viewed from their political power resource possession, the two diametrical conflicts between the two groups of actors were imbalance. On the one hand, the pro-liberalization actors had been controlling stronger political power resources than anti-liberalization actors. The rice traders and distributors had their own financial resources, modern rice milling units, paddy dryer and warehouses for collecting paddy and rice, and they built political and business alliances with bureaucrats, politicians and rice importers, whereas importers, exporters and BULOG had been closely attached with the old New Order regime political oligarchy and also political and business alliance with bureaucrats and pragmatic politicians. The middlemen had been the free rider in the sense that they gained free benefits from collecting paddy and rice channelled to the rice traders and distributors without high financial costs. On the other hand, the peasant villagers and the poor consumers were seriously hit because of the limited political power resources to influence policymaking process. The volume of domestic rice and paddy government procurement had been decreased by 50% since the era of political reform regime, so that the government didn't have strong capacity to combat the rice price speculators. And the government subsidy of fertilizers to the small peasant villagers had been mostly diverted to the fertilizer business actors, bureaucrats and politician because the distribution of fertilizers was mismanaged by business and political actors gaining profit from the mode of fertilizer distribution.[29]

In the era of political reform, political parties, bureaucrats, technocrats and business actors had been working together for the pursuit of financial resources guided by corruption.[30] The high tension of political competition among the political parties needing greater financial support for maintaining their careers compelled them to pursue rent-seeking practices in government bureaucracy at the cost of their public service responsibility. To maintain the domestic market price of rice at a steadily high level to give more incentives to the exporters, the rice distributors had employed an oligopolistic system.[31] Limited by the government regulation,[32] BULOG was not allowed to buy rice and paddy above the price level of HPP, providing an opportunity for the middle-class rice traders and high-class

[29]Husodo, Siswono Yudo, 2005, *Interview;* Basirun, 2011, *Interview.*

[30]Astuti, Dwi, 2011: *Interview.*

[31]Paasch, Amin, op-cit, 2007, p.108.

[32]*Kompas*, 23/3/2007.

distributors to accumulate paddy and rice in their own warehouses and sell them to the consumers bit by bit to preserve consumers' high price rice. The rice distributors located in the big cities did not only sell the rice collected from the middle-class millers but also processed their own paddy collected directly from the peasants by using their own advanced technology milling units, in order to receive higher benefits.[33]

The upper-class millers not only obtained the production input of paddy and rice from domestic peasants but also from rice producers beyond the borders of Indonesia. Such upper-class distributors could be found in Java, Sumatera and Sulawesi islands.[34] The big rice traders having close business relations with Suharto powers circle also obtained an import tender from the government.[35] Considering the negative phenomena, the SBY government should have reformed the domestic commercial management of paddy and rice to reduce price disparity between paddy and rice. However, the SBY government preferred to revise the import ban policy instituted by the Megawati government. Consequently, the rice import escalated.

17.5 Conclusion

Based on the facts, it can be concluded that the multiparty system adopted in Indonesia had been an obstacle for interest groups to take part in the political participation because civil society in Indonesia was still emerging due to Suharto regime restriction and therefore lacked political capacity to influence the policymaking process and its implementation.

The Megawati government supported by reformist political parties, peasant organizations and NGOs implemented de-liberalization policy in the rice sector by imposing rice import tariff, banning import and domestic rice market price stabilization. However, this policy failed to materialize because the political regime was replaced by the newly elected coalition government which was supported by business and pragmatic political parties.

The failure of the rice policy de-liberalization in the era of SBY government was due to the dominant position of the political and business actors linked to the old political oligarchy in rice policymaking process, GOLKAR and its splinter party, DEMOKRAT Party, HANURA and later GERINDRA supported by the pragmatist political parties, IFIs such as IMF and World Bank and the Western countries. Peasant organizations and NGOs involved in policymaking process did not have adequate political power to influence public policy because the pragmatic political

[33]Dinata, Suherman, 2011: *Interview*.

[34]Sukidi, Nellys, 2011: *Interview*.

[35]*Kompas*, 8/9/1998.

parties were reluctant about defending the interests of peasant producers and poor people consumers.

HPP stipulated by the government was not effective in sustaining the price of rice and paddy on the level of peasant producers. The reduction of the scale of paddy and rice government procurement could not be used as an instrument to stabilize the domestic price of paddy and rice, because the rice and paddy produced by the peasant villagers were mostly stored by the big rice traders and distributors. As result of the elimination of LKBI and the reduction of the KUD function, the incidents of the paddy and rice price drops at the level of peasant producers occurred frequently, causing the rice and paddy price at the level of millers and peasant producers to remain mostly under the price level of HPP. Since the end of the New Order regime until the era of SBY reform regime, the level price of GKG was persistently below the HPP price level indicating that the level price of HPP was not effective in sustaining the price stability of paddy and rice.

The domestic price stabilization of paddy and rice was conducted in a limited manner without actually managing to control the big rice traders and distributors with the capacity to create the power of oligopoly in the domestic rice and paddy market. The escalation of price disparity between paddy and rice, where the price of paddy was frequently dropped while the price of rice was sky rocketing, indicated that the business transaction of paddy and rice only benefited big rice traders and distributors. The higher price of international rice compared with domestic rice due to domestic market price manipulation by domestic rice traders and distributors resulted in the speeding up of rice import volume. Consequently, the peasant producers in the villages continued to be increasingly marginalized by the oligopolistic market of paddy and rice, and the rice import surges.

Chapter 18
Identification of Phenotype and the Origin of Rambon Cattle of Indonesia

Yudi Adinata, Aryogi, Dicky Pamungkas, and Tety Hartatik

Abstract The aim of this study was to identify the diversity of local cattle in Indonesia. We analyzed the phenotype of Rambon cattle which is a result of crossing between *Bos sondaicus* (Bali cattle) and *Bos indicus*. The observation data of exterior characteristics (quantitative) was collected from 101 Bali cattle, 101 Madura cattle, and 54 Rambon cattle. Sixteen morphometric traits were measured on each animal. The parts measured were body length (BL), withers height (WH), rump height (RH), rump length (RL), rump width (RW), heart girth (HG), shoulder width (SW), chest depth (CD), face length (FL), face width (FW), horn length (HL), horn diameter (HD), tail length (TL), tail length with hair (TLH), ear length (EL), and ear width (EW). The sequences of mtDNA cytochrome b (cyt b) gene were analyzed from ten samples of each breed to identify the origin of those cattle compared with *Bos taurus* (Limousin). This result showed that for the morphostructural differences between Bali, Madura, and Rambon cattle, significant differences were found in all body measurements. We found two clusters exist, one formed by the Bali and Madura breeds and the other by the Rambon breed. We identified the 32 single nucleotide polymorphisms of cyt b gene. Bali and Rambon cattle have the same haplotype. Mostly Madura cattle (80%) have a different haplotype compared with both cattle. Thus, the result clearly shows the diversity of local cattle in Indonesia and based on the molecular analysis suggests that Rambon cattle have their origin maternally from Bali cattle.

Y. Adinata (✉) • Aryogi • D. Pamungkas
Beef Cattle Research Station, Pasuruan, Jawa Timur 67184, Indonesia
e-mail: adinatayudi@yahoo.com

T. Hartatik
Laboratory of Animal Breeding, Faculty of Animal Science, Gadjah Mada University, Yogyakarta 55281, Indonesia

B. McLellan (ed.), *Sustainable Future for Human Security*,
https://doi.org/10.1007/978-981-10-5430-3_18

18.1 Introduction

The wide range of breeds and species that have evolved in various environments represent unique sets of genetic diversity. Genetic diversity has been defined as the variety of alleles and genotypes present in a population, and this is reflected in morphological, physiological, and behavioral differences between individuals and populations (Frankham et al. 2002). Characterization of a breed of livestock is the first approach to a sustainable use of its animal genetic resource (Lanari et al. 2003). Body measurements have been used in large animals to contrast variation in size and shape (Gilbert et al. 1993; Lanari et al. 2003). These body dimensions have also been used at various times for the estimation of weights when live weights are measured alongside these parameters (Shahin et al. 1995; Yakubu et al. 2005).

The first step in the characterization of local genetic resources is based on the knowledge of variation in the morphological traits (Delgado et al. 2001). Many researchers gave an attention to the body size of livestock animal as the object being observed. Several studies reported a strong correlation between some linear body sizes with some production traits (Otoikhian et al. 2008; Abdel-Moneim 2009). The outcome of genetic improvement programs could also be evaluated on morphological basis (Riva et al. 2004). There are variations in body dimensions of farm animals according to breed type, and one of the ways of differentiating breeds is to evaluate their morphostructural characteristics (Metta et al. 2004). Body measurements are also important in giving information about the morphological structure and development ability of animals (Pesmen and Yardimen 2008). Although recent analyses have focused on molecular techniques, morphometric measurements have been used to evaluate the characteristics of various breeds of animals and could provide useful information on the suitability of animals for selection (Yakubu 2010). The characterization of the genetic structure of populations, breeds, and species provides information necessary for developing breeding program strategies, as it allows genetic variability studies, information that is also important for genetic conservation programs (Thieu et al. 2010).

Molecular markers are a tool to study the diversity on the genetic level. The most widespread use of molecular markers in this context is the assessment of diversity within and between breeds (Baumung et al. 2004). Molecular genetic studies within and across breeds are essential for the effective management of animal genetic resources (Ruane 2000; Simianer 2005). The cytochrome b (cyt b) gene contains abundant phylogenetic information among intra- and interspecies, and it is considered to be a good marker to study the genetic differentiation and phylogenetic relationships among species within the same genus or the same family (Zardoya and Meyer 1996). Cyt b gene is widely used in studies on origin, taxonomy, and phylogeny of the subfamily (Birungi and Arctander 2001).

18.2 Materials and Methods

18.2.1 Collecting Data of Morphostructural Traits

The experiment made use of a random sample of 256 cattle of female sex, comprising 101 Bali, 101 Madura, and 54 Rambon, respectively. The animals were more than 4 years old as determined by dentition. They were reared through the semi-intensive management system and originated from different herds sampled in Bali Island, Madura Island, and Banyuwangi Region in East Java, Indonesia. Efforts were made to restrict sampling to phenotypically pure Bali, Madura, and Rambon cattle, respectively, by measuring only those that conformed to the classification descriptors of third breeds.

Sixteen morphometric traits were measured on each animal. The parts measured were body length (BL), measured from the distance from the point of the shoulder joint to the point of the pin bone; withers height (WH), the distance from the highest point of wither to the ground; rump height (RH), the distance from the highest point of rump to the ground; rump length (RL), rump width (RW), and width of hips (*Tuber coxae*), measured from the hips (*Tuber coxae*) to pins (*Tuber ischii*); heart girth (HG), measured as body circumference just behind the forelegs; shoulder width (SW), measured as the distance from the left to right upper arm (pars cranialis of the tuberculum majus humeri); chest depth (CD), the distance from the highest point of wither to the chest bone just behind the forelegs; face length (FL), the distance from between the horn site to the lower lip; face width (FW), the distance between front of both eyes; horn length (HL), the distance from point of horn attachment to the tip of the horn; horn diameter (HD), the circumference of horn at base; tail length (TL), measured from the tail drop to the tip of the tail; tail length with hair of tail (TLH), measured from the tail drop to the tip of the tail including hair; ear length (EL), distance from the point of attachment to the tip of the ear; and ear width (EW), the circumference of ear at the mid ear. The height measurement (cm) was done using a graduated measuring stick. To achieve this, animals were placed on a flat ground and held by two field assistants. The length and circumference measurements (cm) were effected using a tape rule, while the width measurements (cm) were taken using a calibrated wooden caliper. All measurements were carried out by the same person in order to avoid between-individual variations.

18.2.2 Blood Samples and DNA Extraction

Thirty blood samples were collected from all animals by the jugular vein, and these were saved in tubes containing K_3EDTA and stored at −20 °C until DNA extraction. Genomic DNA was isolated from whole blood by using standard SDS/proteinase K method according to Sambrook et al. (1989). 200 μl of whole blood was moved to a 1.5 ml tube and 800 μl buffer A added; then it was centrifuged at 6000 rpm for 5 minutes. Carefully discarding the upper liquid, an additional 300 µl buffer A was added, and then the process was repeated again centrifuging at 6000 rpm for 5 minutes. Carefully discarding the upper liquid and adding 270 µl buffer B and 30 µl buffer C, the samples were then incubated overnight at 50 °C in a multi-heater. 71 µl of saturated 5 M NaCl was then added and the mixture strongly shaken for 15 sec and then centrifuged at 6000 rpm for 10 minutes, before being taken out and placing the upper liquid in a new tube. The DNA was precipitated using 2-volume pure EtOH at room temperature. The samples were then centrifuged at 12,000 rpm for 10 minutes, and then the upper liquid was discarded. The DNA was washed using 70% EtOH and centrifuged at 12,000 rpm for 5 minutes. The upper liquid was discarded, and the sample was dried at room temperature and dissolved in 100 µl TE buffer or sterile Aqua Bidest. The extracted DNA samples were stored at −20 °C and used later as a substrate for PCR reaction.

18.2.3 Polymerase Chain Reaction (PCR)

The 464-bp fragment of cytochrome b (cyt b) gene was amplified using primers: forward primer of L14735 (5′-AAA AAC CAC CGT TGT TAT TCA ACT-3′) and reverse primer of H15149 (5′-GCC CCT CAG AAT GAT ATT TGT CCT CA-3′) (Wolf et al. 1999). DNA was amplified in a total volume of 20 µl containing 1 µl genomic DNA (10–100 ng), 1 µl each primers, 10 µl PCR KIT (KAPPA2G™ Fast, KAPABIOSYSTEMS, USA), and 7 µl sterile Aqua Bidest. PCR conditions were 2 min at 94 °C, 36 sec at 95 °C, 73 sec at 51 °C, 84 sec at 72 °C, 35 cycles, and 3 min at 72 °C [9]. The PCR was carried out in Primus-25 Advanced (Germany) Thermal Cycler. The PCR products were visualized on 1% agarose gel buffered with 1X Tris-Boric-EDTA buffer (1XTBE), stained with ethidium bromide and visualized under UV light. Ten samples of each breed (Bali, Madura, and Rambon cattle) were sequenced in order to understand the single nucleotide polymorphism (SNP). Total volume 30 µl of PCR product from each sample was sent to 1st Base Genetika Science for further sequencing.

18.2.4 Data Analysis

The morphological traits were subjected to analysis using descriptive statistics and hierarchical clustering using IBM SPSS Statistic 20. The sequence results were analyzed using BioEdit 7.7.

Table 18.1 Descriptive statistics of morphological traits of Bali, Madura, and Rambon cattle

		Bali		Madura		Rambon	
No	Traits	Mean ± SE	SD	Mean ± SE	SD	Mean ± SE	SD
1.	BL	113,82 ± 0,47[a]	4,75	130,46 ± 0,61[b]	6,12	108,59 ± 1,06[c]	7,50
2.	WH	114,76 ± 0,37[a]	3,69	116,00 ± 0,61[b]	6,13	119,51 ± 1,10[c]	8,00
3.	RH	114,90 ± 0,44[a]	4,38	118,60 ± 0,55[b]	5,58	123,08 ± 1,00[c]	7,20
4.	RL	37,23 ± 0,22[a]	2,25	34,67 ± 0,32[b]	3,22	30,80 ± 0,38[c]	2,75
5.	RW	34,66 ± 0,23[a]	2,29	25,23 ± 0,27[b]	2,76	39,50 ± 0,44[c]	3,09
6.	HG	157,82 ± 0,72[a]	7,27	144,90 ± 1,08[b]	10,82	163,34 ± 1,60[c]	11,54
7.	SW	31,56 ± 0,27[a]	2,68	30,80 ± 0,33[b]	3,27	33,12 ± 0,65[c]	4,66
8.	CD	61,70 ± 0,40[a]	4,03	55,86 ± 0,37[b]	3,76	60,79 ± 0,69[c]	4,98
9.	FL	36,20 ± 0,15[a]	1,48	40,14 ± 0,18[b]	1,82	40,56 ± 0,38[c]	2,74
10.	FW	16,83 ± 0,12[a]	1,17	18,04 ± 0,13[b]	1,27	18,21 ± 0,17[c]	1,23
11.	HL	17,59 ± 0,38[a]	3,82	8,37 ± 0,28[b]	2,82	23,07 ± 0,76[c]	5,50
12.	HD	15,04 ± 0,15[a]	1,51	11,49 ± 0,20[b]	1,96	15,00 ± 0,31[c]	2,24
13.	TL	69,26 ± 0,45[a]	4,55	80,96 ± 0,61[b]	6,12	82,53 ± 0,87[c]	6,19
14.	TLH	93,03 ± 0,66[a]	6,64	100,65 ± 0,86[b]	8,67	109,14 ± 1,16[c]	8,31
15.	EL	23,00 ± 0,17[a]	1,71	20,71 ± 0,15[b]	1,51	23,27 ± 0,29[c]	2,11
16	EW	13,96 ± 0,08[a]	0,83	12,54 ± 0,09[b]	0,90	15,34 ± 0,27[c]	1,99

SE standard error, *SD* standard deviation
Means in the same row with different superscripts are significantly different ($p < 0.05$)

18.3 Results and Discussion

18.3.1 Phenotype Base on the Morphostructural Traits

Table 18.1 shows the mean, standard error, and standard deviation for each of the morphometric characters of the Bali, Madura, and Rambon cattle investigated. Bali, Madura, and Rambon cattle had significant ($p < 0.05$) different mean values for all the body measurements estimated. In the morphostructural differences between Bali, Madura, and Rambon cattle, significant differences were found in all body measurements. It can be pointed out that in Bali, Madura, and Rambon cattle, there are differences in body dimensions; the study corresponds to a different cattle format. The differences in measurements in the Bali, Madura, and Rambon cattle populations could be as a result of the influence of the environment and peculiarities of the breed. Morphology expresses a strong relationship with productive potential, since it contains the structure which supports the biological functionality of the animal (Alpak et al. 2009). Morphostructure provides information susceptible to be used on ethnological characterization of an animal population and allows a judgment of the productive potential based on the implicit mechanical relationships within the morphologic structure (Yakubu et al. 2010). If this relationship is not considered, it would imply that appropriate productive life adaptation models would not be right (Alpak et al. 2009).

Facial and head differences are very important from an ethnological point of view in breed identification; the ethnological differences are the basic concepts on the foundation of a breed (Sierra Alfranca 2001). The body measurements are related to meat value (rump length, chest width, heart girth). These morphostructural changes are related to changes in the productive performance of the animal format; therefore, in the sustained selection process, not only are changes in ethnological traits involved but also productive traits (Yakubu 2010; Yakubu et al. 2010). Performance, especially with regard to meat production, can be assessed from body measurements that are less closely associated with bone growth.

The measurements of rump length, chest width, and heart girth of Rambon cattle were higher than Bali and Madura cattle. Body weights of Rambon, Bali, and Madura cattle were 347,34 kg, 262,32 kg, and 237,81 kg. Based on these measurements, it was indicated that the animal format of Rambon cattle had a body frame which can support higher meat production.

18.3.2 Phylogenetic Tree Analysis

The morphometric cluster analysis and dendrogram using morphostructural variables were shown in Table 18.2 and Fig. 18.1. It appears that three breeds of cattle can be differentiated. Subpopulations can be formed due to limited gene flow and geographical isolation so that the subpopulations have different phenotypic characteristics. The body form (type, structure, or proportions) is supposed to be ideal for the purpose for which the animal is produced. The morphometric cluster analysis shows that the Bali, Madura, and Rambon cattle populations were quite distinctive breeds.

The Euclidean distance coefficient (D) between the three breeds calculated from the 16 body dimensions was found to be as presented in Table 18.2 which shows the distance apart between the three breeds considering their relatedness, and this is suggestive that Bali and Madura cattle are less related genetically with Rambon cattle.

In the dendrogram (Fig. 18.1), two clusters exist, one formed by the Bali and Madura breeds and the other by the Rambon breed. The differences in morphological distance between the three cattle populations coupled with high correct assignment to source genetic groups are an indication that they belong to different breeds. This could have been facilitated by the fact that measurements were restricted to phenotypically pure animals. The use of multivariate discriminant analyses therefore could be successfully used in morphometric differentiation.

Table 18.2 Euclidean distance between the breeds

Breed	1:Bali	2:Madura	3:Rambon
1:Bali	0,000	27,625	28,920
2:Madura	27,625	0,000	39,454
3:Rambon	28,920	39,454	0,000

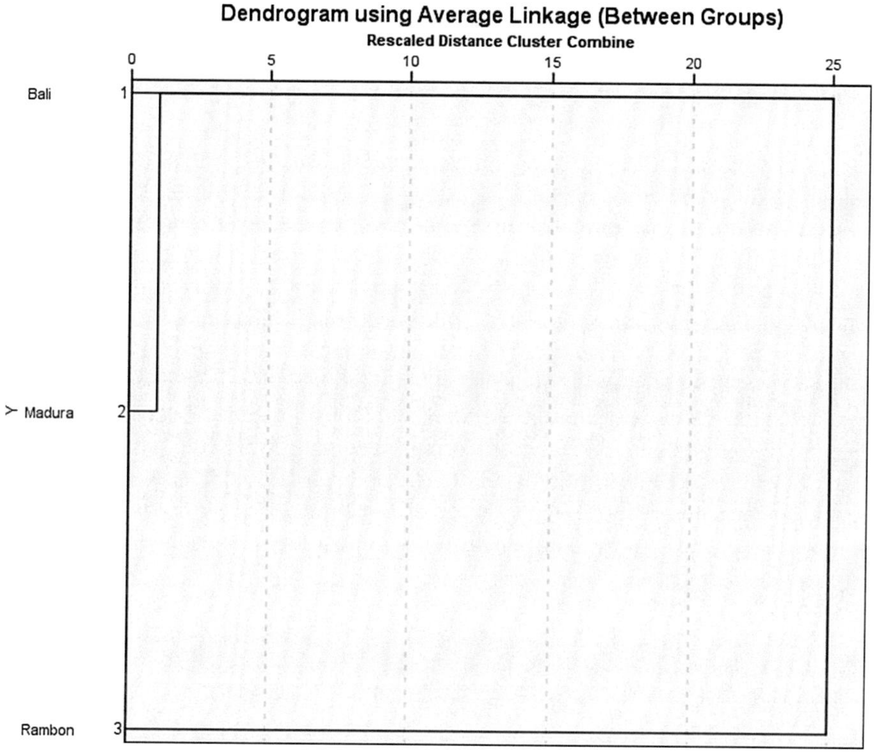

Fig. 18.1 Dendrogram based on the Euclidean distance of the Bali, Madura, and Rambon cattle

18.3.3 Single Nucleotide Polymorphism of Cytochrome b

Thirty samples of DNA from Bali, Madura, and Rambon cattle showed two haplotypes. Bali and Rambon cattle have 100% similar sequence of cytochrome b. However 80% of Madura cattle population have distinct sequences of cyt b compared with those of Bali and Rambon cattle. We found 32 single nucleotide polymorphism of cyt b gen in eight Madura cattle. Two out of ten samples in Madura cattle population were similar with Bali and Rambon cattle. Thus, Madura cattle have diversity in the population.

Based on the molecular analysis, there were indications that Rambon cattle have the same origin as Bali cattle. In line with the morphometric study, Bali and Madura cattle were in the same cluster, and Rambon cattle was in another. It can be explained that the Bali cattle pure breed can be used to form another crossbreed. Therefore it is needed to conserve and develop the population of pure breed Bali cattle. Thus, breeding management to keep the sustainability of Bali Cattle population to obtain genetic resources for crossbreeding is important to obtain cattle with optimum meat production trait.

18.4 Conclusion

In the morphostructural differences between Bali, Madura, and Rambon cattle, significant differences were found in all body measurements. In the dendrogram, two clusters exist, one formed by the Bali and Madura breeds and the other by the Rambon breed. The use of multivariate analysis provided means for objective description of livestock breed using body measurements. The morphometric cluster analysis and dendrogram differentiate the three breeds morphologically into distinctive genetic breeds. The present information on the morphometric differentiation of Bali, Madura, and Rambon breeds of cattle could be complemented with genetic characterization using biochemical and DNA markers. Molecular analysis suggests that Rambon cattle have the origin maternally from Bali cattle. Madura cattle have the diversity in the population. This could aid field assessment, management, and conservation of the three cattle populations, where the goal is to obtain phenotypically pure local genetic resources for future selection and breeding improvement strategies.

Acknowledgment This research was supported by BCATRES Indonesia. Thank you to all the staff at the Laboratory of Animal Breeding, Faculty of Animal Science, who supported the facilities for the analysis of DNA. We also would like to S. D. Volkandari for technical assistance for DNA extraction.

References

Abdel-Moneim AY (2009) Use of live body measurements for prediction of body and carcass cuts weights in three Egyptian breeds of sheep. Egyptian J Sheep and Goat Sci 2:17–32

Alpak H, Onar V, Mutus R (2009) The relationship between morphometric and long bone measurements of the Morkaraman sheep. Turk J Vet Anim Sci 33(3):199–207

Baumung R, Simianer H, Hoffmann I (2004) Genetic diversity studies in farm animals – a survey. J Anim Breed Genet 121:361–373

Birungi J, Arctander P (2001) Molecular systematics and phylogeny of the Reduncini (Artiodactyla:Bovidae) inferred from the analysis of mitochondrial cytochrome B gene sequences. J Mamm Evol 8:125–147

Delgado JV, Barba C, Camacho ME, Sereno FTPS, Martinez A, Vega-Pla JL (2001) Livestock characterisation in Spain. Anim Genet Resour Inform 29:7–18

Frankham R, Ballou JD, Briscoe DA (2002) Introduction to conservative genetics. Cambridge University Press

Gilbert RP, Bailey DRC, Shannon NH (1993) Linear body measurements of cattle before and after 20 years of selection for post-weaning gain when fed two different diets. J Anim Sci 71:1712–1720

Lanari MR, Taddeo H, Domingo E, Centeno MP, Gallo L (2003) Phenotypic differentiation of exterior traits in local Criollo goat population in Patagonia (Argentina). Arch TierzDummerstorf 46:347–356

Metta M, Kanginakudru S, Gudiseva N, Nagaraja J (2004) Genetic characterization of the Indian cattle breeds, Ongole and Deoni (Bos indicus), using microsatellite markers – a preliminary study. BMC Genet 2004 5:16

Otoikhian CSO, Otoikhian AM, Akporhuarho OP, Isidahomen C (2008) Correlation of body weight and some body measurement parameters in Ouda sheep under extensive management system. Afr J Gen Agric 3:129–133

Pesmen G, Yardimen M (2008) Estimating the live weight using some measurements in Saanen goats. Archive Zoo Technical 11(4):30–40

Riva J, Rizzi R, Marelli S, Cavalchini LG (2004) Body measurements in Bergamasca sheep. Small Rumin Res 55:221–227

Ruane J (2000) A framework for prioritizing domestic animal breeds for conservation purposes at the national level: a Norwegian case study. Conserv Biol 14:1385–1393

Sambrook J, Fritsch EF, Maniatis T (1989) Molecular cloning, a laboratory manual. Cold Spring Harbour Laboratory Press, Cold Spring Harbour

Shahin KA, Soliman AM, Moukhtar AE (1995) Sources of shared variability for the Egyptian cattle body shape (conformation). Indian J Anim Sci 65:759–764

Sierra Alfranca I (2001) The breed concept: evolution and reality. Arch Zootec 50:547–564

Simianer H (2005) Decision making in livestock conservation. Ecol Econ 53:559–572

Thieu PX, Georgescu SE, Manea MA, Hermenean AO, Costache M (2010) Genetic diversity and phylogenetic relationships of Romanian cattle breeds inferred from cytochrome b gene partial sequences. Romanian Biotechnol Lett 15(2)

Wolf C, Rentsch J, Hübner P (1999) PCR-RFLP analysis of mitochondrial DNA: a reliable method for species identification. J Agric Food Chem 47(4):1350–1355

Yakubu A (2010) Path coefficient and path analysis of body weight and biometric traits in Yankasa lambs. Slovak J Anim Sci 43:17–25

Yakubu A, Salako AE, Ogah M, Ari MM (2005) Relationship between body weight and chest girth in white Fulani cattle. J Nat Appl Sci 1:17 21

Yakubu A, Salako AE, Imumorin IG, Ige AO, Akinyemi MO (2010) Discriminant analysis of morphometric differentiation in the west African dwarf and red Sokoto goats. S Afr J Anim Sci 40:381–387

Zardoya R, Meyer A (1996) Phylogenetic performance mitochondrial protein coding genes in resolving relationship among vertebrate. Mol Biol Evol 13:933–942

Chapter 19
Machine Vision-Based Analysis for Black Tea Quality Evaluation

Avicienna Ulhaq Muqodas, Slamet Widodo, Kudang B. Seminar, and Mohamad Solahudin

Abstract Currently, visual quality evaluation of black tea production depends only on expert labor judgement. This allows biases and inconsistency in the end product caused by unstandardized methods. Therefore, it is proposed that machine vision could be applied in order to standardize the quality evaluation method. The goal of this research is to build a machine vision-based system to classify quality classes of black tea. In this research black tea classes used are class I, class II, and class III without considering the grade for each class. A neural network classifier was used to classify black tea images with seven nodes, one hidden layer. Ten features were used as an input for the classifier, RGB, index RGB, R/G, R/B, magenta, and yellow. From the results, the system achieved an accuracy of 100%. Application of machine vision technology in black tea production could not only support sustainability in agriculture, but also can also be used for precision agriculture in practical fields.

19.1 Introduction

Sustainability issues in agriculture are not only about on-farm operations, but off-farm operations are also noteworthy. Poor off-farm handling can impact on the success or failure of larger agricultural activities. This means that agricultural activities become unsustainable. Therefore, maintaining quality of products from the agricultural sector is essential, and studies that support off-farm handling are critical to ensuring sustainability in agriculture (Juansah et al. 2014).

Nowadays, black tea quality evaluation according to the Indonesia national standard, is measured using a number of parameters, but the visual appearance of black tea is the most common and simplest method to determine the quality of black tea in the production line. There are a number of parameters that are measured: particle shape, color, cleanliness, and tip.

A.U. Muqodas • S. Widodo • K.B. Seminar (✉) • M. Solahudin
Department of Mechanical and Biosystem Engineering, Bogor Agricultural University, Kampus IPB Darmaga, 220, Bogor, West Java 16680, Indonesia
e-mail: seminarkudangboro@gmail.com

B. McLellan (ed.), *Sustainable Future for Human Security*,
https://doi.org/10.1007/978-981-10-5430-3_19

Currently, this visual appearance method only depends on expert labor judgement, which can cause biases and inconsistency of end product caused by the unstandardized method of evaluation. This inconsistency makes it difficult to establish quality standards. For producers, quality standards are not fixed, causing fluctuating product quality and can cause consumer distrust. In addition, by relying on a specific person, inheritance of ability becomes difficult.

Machine vision is the technology and method used to provide imaging-based automatic inspection and analysis for such applications as automatic inspection, process control, and robot guidance in industry (Graves and Batchelor 2003). It has been shown that using imaging technology and automated visual inspection for sensing agricultural product is accurate, low cost, high speed, and nondestructive and yields consistent result (Chen et al. 2002; Brosnan and Sun 2004). Applying machine vision technology in the production of black tea makes it possible to develop a better quality assessment method. This study not only supports sustainability in agriculture but can also be seen as a proof of concept of precision agriculture (Sampurno et al. 2014; Heriyanto et al. 2016) in a practical field.

19.2 Method

In this research, black tea was obtained from a research institute for tea and cinchona, Gambung, West Java. According to Rohdiana et al. (2014) the quality classes in black tea classification are class I, class II, and class III. In this research quality classes were used without considering the grade of each class.

Image Acquisition In this research images were taken using a digital camera Panasonic LUMIX DMC-F5 in chamber with two light sources of LED lamps. Illustration and work set configuration of image acquisition can be seen in Table 19.1 and Fig. 19.1. The images acquired in this research totaled 154 images with 65 images of class I, 44 images of class II, and 45 images of class III. Examples of the images are shown in Fig. 19.2.

Table 19.1 Work set configuration

Parameter	Explanation
Camera height	10 cm
Lamp height	12.5 cm
Picture resolution (pixel)	2048 × 1536
Lamp properties	2 LED 5 Watt
Light intensity	450 lux

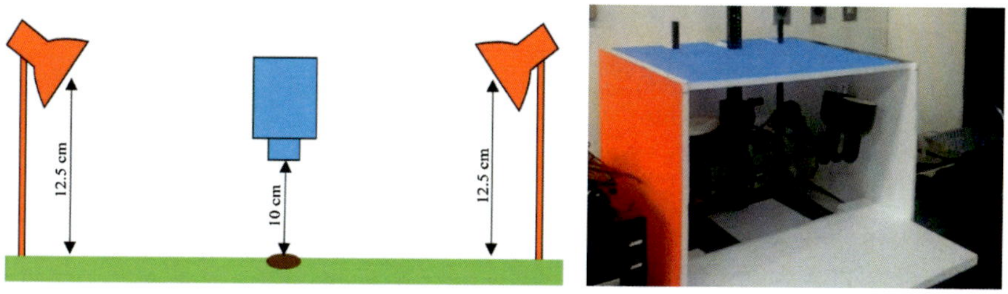

Fig. 19.1 *Left*, work set illustration; *right*, image acquisition chamber

Fig. 19.2 Example of acquired image: (**a**) class I; (**b**) class II; (**c**) class III

Color Feature RGB color space is a basic feature in computer images that could be extracted without any conversion. RGB corresponds to the three basic colors: red, green, and blue, respectively. In this research seven more features are generated from basic RGB features; these were RGB index, yellow, magenta, red/green ratio, and red/blue ratio. Each feature on every image was averaged. Color features can be described as following equations:

$$\text{Average red value} = \frac{1}{n}\sum_{i=1}^{M} R$$

$$\text{Average green value} = \frac{1}{n}\sum_{i=1}^{M} G$$

$$\text{Average blue value} = \frac{1}{n}\sum_{i=1}^{M} B$$

$$\text{Average red/blue ratio} = \frac{1}{n}\sum_{i=1}^{M} \frac{R}{B}$$

$$\text{Average red/green ratio} = \frac{1}{n}\sum_{i=1}^{M} \frac{R}{G}$$

$$\text{Average red index} = \frac{1}{n}\sum_{i=1}^{M}\frac{R}{R+G+B}$$

$$\text{Average green index} = \frac{1}{n}\sum_{i=1}^{M}\frac{G}{R+G+B}$$

$$\text{Average blue index} = \frac{1}{n}\sum_{i=1}^{M}\frac{B}{R+G+B}$$

where R is the red value, G green value, B blue value, and n the total number of pixels.

The yellow and magenta color feature from CMYK color space is also involved in this research. CMYK color space is usually utilized in the printing process. CMYK is a subtractive color space that involves reflected light and colorants. In CMYK, the reflective substrate is assumed to be white.

$$\text{Average yellow value} = \frac{1}{n}\sum_{i=1}^{M}\left[\frac{((1-Y)-\text{Min}(C,M,Y))}{(1-\text{Min}(C,M,Y))}\right]$$

$$\text{Average magenta value} = \frac{1}{n}\sum_{i=1}^{M}\left[\frac{((1-M)-\text{Min}(C,M,Y))}{(1-\text{Min}(C,M,Y))}\right]$$

Where

$$C = 1-(R/255),$$
$$M = 1-(G/255),$$
$$Y = 1-(B/255)$$

Neural Network for Classification A neural network is one pattern recognition technique that is feasible for solving nonlinear problems. In the image processing field, neural networks are commonly used not only for object recognition but also for preprocessing, data reduction, segmentation, and image understanding (Egmont-Petersen et al. 2002).

In this research a neural network is conducted as an algorithm to classify the instances of black tea image. The neural network was constructed with ten nodes in the input layer, one hidden layer with seven nodes, and a three node output layer. The image of the neural network architecture is shown in Fig. 19.3.

Validation Method The K-fold cross validation method was conducted to avoid prediction bias in this research. The method in the k-fold cross validation method is to split the data into k sets. $K-1$ of the data is used as training data, and the $k-(k-1)$ is used as test data. This process is repeated until training and testing is completed. Overall accuracy is calculated with the following expression:

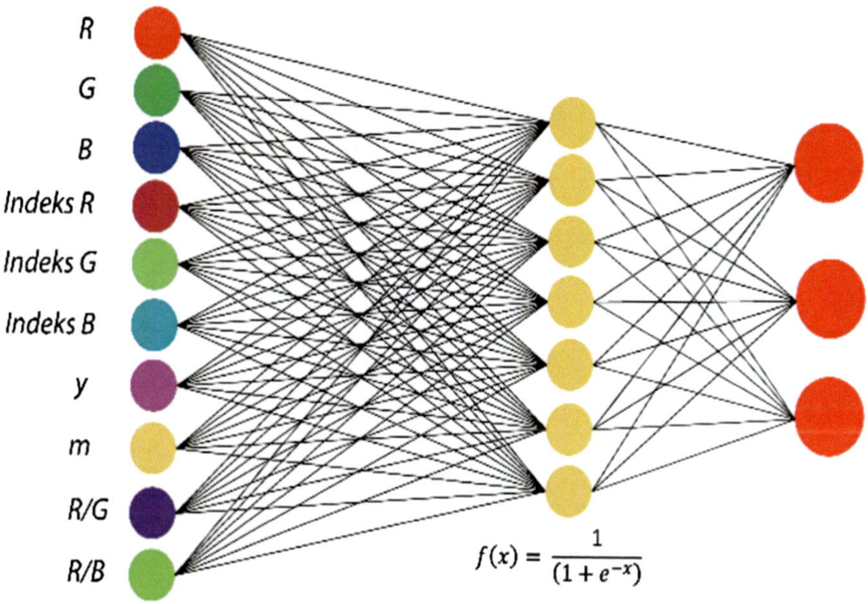

Fig. 19.3 Neural network structure diagram

$$\text{Accuracy} = \frac{1}{k}\sum_{i=1}^{k} A_i$$

Where A_i, is accuracy for each validation loop.

To ensure the classification algorithm gives reliable results, k-cross validation method is used to validate the algorithm, with number of $k = 5$.

19.3 Results

Feature Value from Sample Image Classification is easily done if separable regions exist in the current data. To ensure the existence of separable regions, each feature was plotted into individual value graphs as shown in Fig. 19.4.

From the graphs above, separable regions exists for some features, although other feature data showed an overlap. For example, in the index R graph, although some data is misclassified, generally we can separate class I, class II, and class III easily by drawing a dividing line between them as seen in Fig. 19.5. Generally, we can conclude that classification can possibly be conducted using chosen features.

But, as seen in Fig. 19.4, RGB value is the most overlapped feature from the chosen features. The number of features which will be utilized is decided using trial and error experiments. The trial and error experiment is conducted using the neural

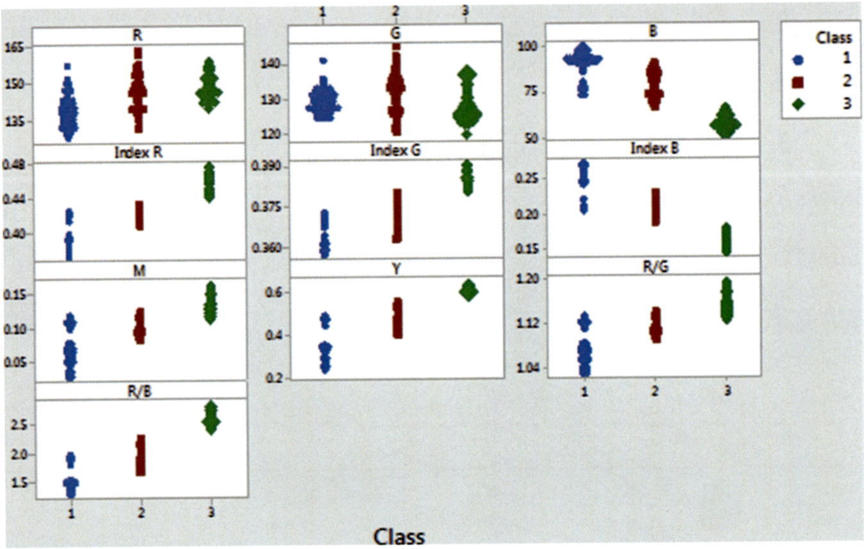

Fig. 19.4 Individual value plot for each feature

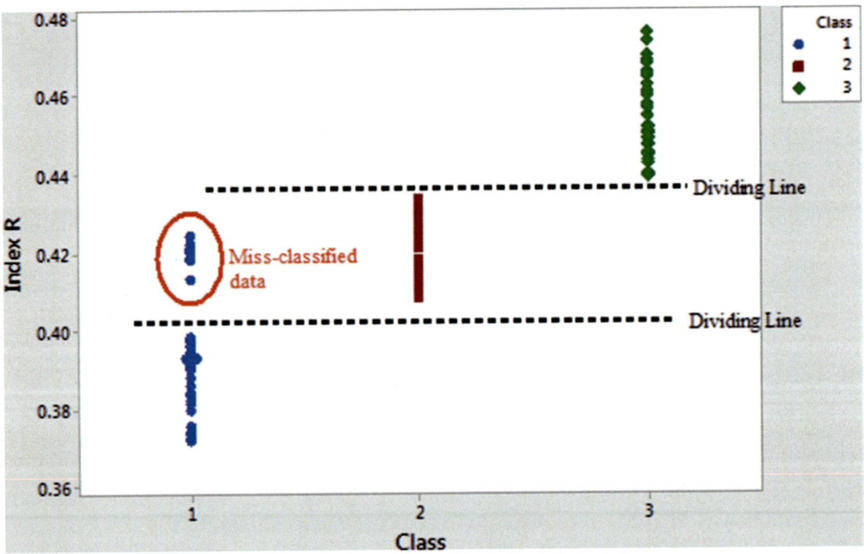

Fig. 19.5 Dividing line and misclassified data in feature example

network classifier with several alternative numbers of features: two features (R/G and R/B feature), three features (index R, index G, and index B), five features (index R, index G, index B, R/G, and R/B), seven features (index R, index G, index B, R/G, R/B, yellow, magenta), and ten features (index R, index G, index B,

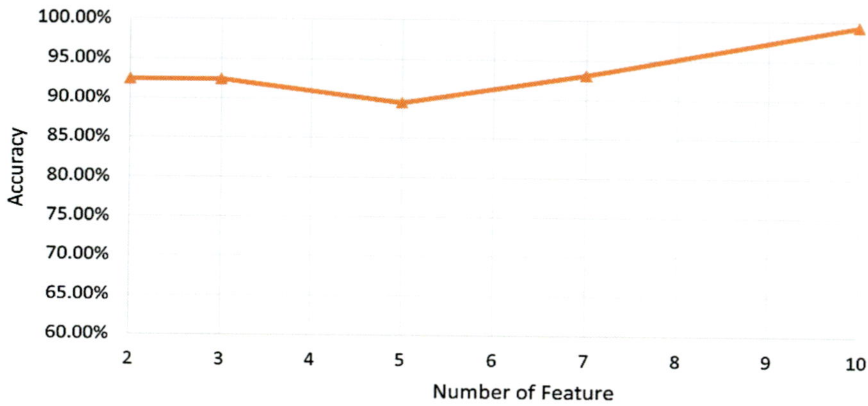

Fig. 19.6 Trial and error experiment using neural network classifier with several number of features

Table 19.2 Confusion matrix of classification result

Ideal	Detected			Correct	Incorrect
	Class I	Class II	Class III		
Class I	65	0	0	100%	0%
Class II	0	44	0	100%	0%
Class III	0	0	45	100%	0%
Total				154	0%
Total accuracy				100%	0%

R/G, R/B, yellow, magenta, red, green, and blue). The trial and error experiment aims not only to prove the features used as inputs generate maximum accuracy but is also used to select the optimal number of features. Accuracy comparison from the trial and error experiment is shown in Fig. 19.6.

Classification Using Neural Network Classifier In this research the neural network was trained using the back propagation method with a learning rate of 0.3 and momentum of 0.2. Training was done using 700 epochs. Neural network training was done using WEKA (Waikato Environment for Knowledge Analysis). Correct prediction for each class is expressed with the confusion matrix shown in Table 19.2.

From the results above, it can be seen that the neural network algorithms can classify class I, class II, and class II perfectly. This is caused by the selection of the best input features to be utilized. The best input may be defined as an input which has a clear dividing area, so that the main function expected of the neural networks is to minimize misclassification in grade estimation.

19.4 Conclusion

The machine learning-based black tea quality class evaluation system was developed very well. The evaluation system was able to identify black tea quality classes using a neural network classifier with 100% accuracy. This means that the system classifies perfectly the black tea classes without any misclassification result.

As future work, it is planned to apply the system in a specific device which will completely analyze all of the visual quality aspects of black tea. To enhance accuracy and processing speed of the developed sorting machine, the use of a higher resolution camera and the modification of neural network model for image recognition will be examined.

References

Brosnan Tadhg, Da-Wen Sun (2004) Improving quality inspection of food products by computer vision—a review. J Food Eng 61(1):3–16

Chen Yud-Ren, Kuanglin Chao, Moon S. Kim (2002) Machine vision technology for agricultural applications. Comput Electron Agric 36(2):173–191

Egmont-Petersen M, de Ridder D, Handels H (2002) Image processing with neural networks—a review. Pattern Recogn 35(10):2279–2301

Graves M, Batchelor B (2003) Machine vision for the inspection of natural products, 2nd edn. Springer, London

Heriyanto H, Seminar KB, Solahudin M et al (2016) Water supply pumping control system using PWM based on precision agriculture principles[J]. IAEJ 25(2):1–8

Juansah J, Budiastra IW, Dahlan K, Seminar KB (2014) Electrical properties of garut citrus fruits at Low alternating current signal and its correlation with physicochemical properties during maturation. Int J Food Prop 17(7):1498–1517

Rohdiana D, Suganda AG, Wirasutsna KR, Iwo MI (2014) Xanthine oxidase inhibitory and immunomodulatory activities of fifteen grades Indonesia orthodox black tea. Int J Pharm Pharm Sci 6:39–42

Sampurno RM, Seminar KB, Suharnoto Y (2014) Weed control decision support system based on precision agriculture approach. TELKOMNIKA (Telecommunication Computing Electronics and Control) 12(2):475

Kolibri 360 GmbH
Ferdinand-Jühlke-Straße 7
99095 Erfurt, Deutschland
produktsicherheit@kolibri360.de